REDUCING RESPONSE BURDEN IN THE AMERICAN COMMUNITY SURVEY

Proceedings of a Workshop

Thomas J. Plewes, *Rapporteur*

Committee on National Statistics

Division of Behavioral and Social Sciences and Education

The National Academies of
SCIENCES · ENGINEERING · MEDICINE

THE NATIONAL ACADEMIES PRESS
Washington, DC
www.nap.edu

THE NATIONAL ACADEMIES PRESS   500 Fifth Street, NW   Washington, DC 20001

This activity was supported by the U.S. Census Bureau through Contract No. YA1323-15-CN-0025. Support of the work of the Committee on National Statistics is provided by a consortium of federal agencies through a grant from the National Science Foundation (No. SES-1024012). Any opinions, findings, conclusions, or recommendations expressed in this publication do not necessarily reflect the views of any organization or agency that provided support for the project.

International Standard Book Number-13:   978-0-309-44943-4
International Standard Book Number-10:   0-309-44943-X
Digital Object Identifier: 10.17226/23639

Additional copies of this report are available from the National Academies Press, 500 Fifth Street, NW, Keck 360, Washington, DC 20001; (800) 624-6242 or (202) 334-3313; http://www.nap.edu.

Copyright 2016 by the National Academy of Sciences. All rights reserved.

Printed in the United States of America.

Suggested citation: National Academies of Sciences, Engineering, and Medicine. (2016). *Reducing Response Burden in the American Community Survey: Proceedings of a Workshop*. Washington, DC: The National Academies Press. doi: 10.17226/23639.

The National Academies of
SCIENCES · ENGINEERING · MEDICINE

The **National Academy of Sciences** was established in 1863 by an Act of Congress, signed by President Lincoln, as a private, nongovernmental institution to advise the nation on issues related to science and technology. Members are elected by their peers for outstanding contributions to research. Dr. Marcia McNutt is president.

The **National Academy of Engineering** was established in 1964 under the charter of the National Academy of Sciences to bring the practices of engineering to advising the nation. Members are elected by their peers for extraordinary contributions to engineering. Dr. C. D. Mote, Jr., is president.

The **National Academy of Medicine** (formerly the Institute of Medicine) was established in 1970 under the charter of the National Academy of Sciences to advise the nation on medical and health issues. Members are elected by their peers for distinguished contributions to medicine and health. Dr. Victor J. Dzau is president.

The three Academies work together as the **National Academies of Sciences, Engineering, and Medicine** to provide independent, objective analysis and advice to the nation and conduct other activities to solve complex problems and inform public policy decisions. The Academies also encourage education and research, recognize outstanding contributions to knowledge, and increase public understanding in matters of science, engineering, and medicine.

Learn more about the National Academies of Sciences, Engineering, and Medicine at **www.national-academies.org**.

The National Academies of
SCIENCES · ENGINEERING · MEDICINE

Reports document the evidence-based consensus of an authoring committee of experts. Reports typically include findings, conclusions, and recommendations based on information gathered by the committee and committee deliberations. Reports are peer reviewed and are approved by the National Academies of Sciences, Engineering, and Medicine.

Proceedings chronicle the presentations and discussions at a workshop, symposium, or other convening event. The statements and opinions contained in proceedings are those of the participants and have not been endorsed by other participants, the planning committee, or the National Academies of Sciences, Engineering, and Medicine.

For information about other products and activities of the National Academies, please visit nationalacademies.org/whatwedo.

STEERING COMMITTEE FOR WORKSHOP ON RESPONDENT BURDEN IN THE AMERICAN COMMUNITY SURVEY

LINDA GAGE (*Cochair*), Demographic Research Unit, Department of Finance, State of California (retired)
JOSEPH SALVO (*Cochair*), Population Division, New York City Department of City Planning
DAVID DOLSON, Social Survey Methods, Statistics Canada
JOHN ELTINGE, Office of Survey Methods Research, Bureau of Labor Statistics
DAVID HUBBLE, Statistical Staff, Westat
JULIA LANE, Wagner Graduate School of Public Service, New York University
NANCY MATHIOWETZ, University of Wisconsin–Milwaukee (emerita)

BRIAN HARRIS-KOJETIN, *Study Director*
MICHAEL COHEN, *Senior Program Officer*
AGNES GASKIN, *Administrative Assistant*

COMMITTEE ON NATIONAL STATISTICS

LAWRENCE D. BROWN (*Chair*), Department of Statistics, The Wharton School, University of Pennsylvania
FRANCINE BLAU, Department of Economics, Cornell University
MARY ELLEN BOCK, Department of Statistics (emerita), Purdue University
MICHAEL CHERNEW, Department of Health Care Policy, Harvard Medical School
JANET CURRIE, Woodrow Wilson School of Public and International Affairs, Princeton University
DONALD DILLMAN, Social and Economic Sciences Research Center, Washington State University
CONSTANTINE GATSONIS, Department of Biostatistics and Center for Statistical Sciences, Brown University
JAMES S. HOUSE, Survey Research Center, Institute for Social Research, University of Michigan
THOMAS MESENBOURG, U.S. Census Bureau (retired)
SUSAN MURPHY, Department of Statistics and Institute for Social Research, University of Michigan
SARAH NUSSER, Office of the Vice President for Research, Iowa State University
COLM O'MUIRCHEARTAIGH, Harris School of Public Policy Studies, University of Chicago
RUTH PETERSON, Criminal Justice Research Center, Ohio State University
ROBERTO RIGOBON, Sloan School of Management, Massachusetts Institute of Technology
EDWARD SHORTLIFFE, Department of Biomedical Informatics, Columbia University and Arizona State University

CONSTANCE F. CITRO, *Director*
BRIAN HARRIS-KOJETIN, *Deputy Director*

Acknowledgments

This workshop was the culmination of an intense period of scoping, planning, and development on the part of the staff of the U.S. Census Bureau and the Committee on National Statistics of the National Academies of Sciences, Engineering, and Medicine, as well as the volunteer members of the workshop steering committee to consider the challenges and opportunities for reducing respondent burden of the American Community Survey (ACS), which is conducted by the U.S. Census Bureau. The 2-day workshop, held in March 2016, included a wide range of experts and stakeholders. It also marked the beginning of a challenging series of expert meetings conducted during the spring of 2016 on various topics pertaining to understanding and controlling for the burden imposed on respondents by the ACS.

The knowledgeable contributions of Census Bureau staff, especially Mark Asiala, Judy Belton, Donna Daily, Todd Hughes, Amy O'Hara, Elizabeth Poehler, David Raglin, Jennifer Reichert, Deborah Stempowski, Anthony Tersine, and Victoria Velkoff throughout this process were very much appreciated.

The workshop was developed with the input and guidance of our exceptionally dedicated and productive fellow members of the steering committee. Selected for their individual expertise on the ACS and the subject matter under consideration, the steering committee met via several telephone conference calls over a 3-month period to design the workshop. The members of the steering committee are commended for their contributions to the enterprise.

These proceedings are the main product of the workshop. This report was prepared by an independent rapporteur whose charter was to distill the gist of the presentations and the essence of the discussions. The steering committee's role was limited to planning and convening the workshop. The views contained in the report are those of individual workshop participants and do not necessarily represent the views of all workshop participants, the planning committee, or the National Academies.

This workshop summary has been reviewed in draft form by individuals chosen for their diverse perspectives and technical expertise, in accordance with procedures approved by the Report Review Committee of the National Academies. The purpose of this independent review is to provide candid and critical comments that will assist the institution in making its published report as sound as possible and to ensure that the report meets institutional standards for objectivity, evidence, and responsiveness to the charge. The review comments and draft manuscript remain confidential to protect the integrity of the process.

We thank the following individuals for their review of this workshop summary: Linda Gage, consultant, Sacramento, CA; Linda A. Jacobsen, U.S. Programs, Population Reference Bureau; Sarah M. Nusser, Center for Survey Statistics and Methodology and Office of the Vice President for Research, Iowa State University; Susan Schechter, senior fellow, NORC at the University of Chicago; and Daniel H. Weinberg, principal, DHW Consulting.

Although the reviewers listed above provided many constructive comments and suggestions, they did not see the final draft of the workshop summary before its release. The review of this report was overseen by Sarah M. Nusser, Center for Survey Statistics and Methodology and Office of the Vice President for Research, Iowa State University. Appointed by the National Academies, she was responsible for making certain that an independent examination of this report was carried out in accordance with institutional procedures and that all review comments were carefully considered. Responsibility for the final content of this report rests entirely with the rapporteur and the institution.

Linda Gage, *Cochair*
Joseph Salvo, *Cochair*
Steering Committee for Workshop on
Respondent Burden in the American Community Survey

Contents

1	Introduction: Understanding Response Burden	1
2	Approaches to Reducing Response Burden	5
3	Improving Response by Building Respondent Support	13
4	Using Administrative Records to Reduce Response Burden	39
5	Using Improved Sampling and Other Methods to Reduce Response Burden	55
6	Tailoring Collection of Information from Group Quarters	79
7	Future Directions	91
References		95

Appendixes

    A Workshop Agenda   99

    B Biographical Sketches of Steering Committee Members and Presenters   105

1

Introduction: Understanding Response Burden

Although people in the United States have historically been reasonably supportive of federal censuses and surveys, they are increasingly unavailable for or not willing to respond to interview requests from federal as well as private sources (National Research Council, 2013b). Moreover, even when people agree to respond to a survey, they increasingly decline to complete all questions, and both survey and item nonresponse are growing problems (National Research Council, 2013b).

In recent years, the American Community Survey (ACS)—the mandatory survey that replaced the census long form that was last used in 2000—has seen an increase in nonresponse, and it has been a target of criticism for invasion of privacy and excessive burden.[1] Although it covers far fewer people than the census long form, it is large by any other measure, requesting responses from 295,000 households every month. The ACS is very visible in the public eye, and it generates a small but continuous stream of complaints to members of Congress, who have held several congressional hearings on the survey.

Four items on the survey have been identified by the Census Bureau as giving rise to the most complaints—income, disability, time of leaving for work, and plumbing facilities (U.S. Census Bureau, 2014). Some of these items are seen by many as intrusive, and the questions attempting to

---

[1]In the decennial censuses from 1960 through 2000, a sample of households (one in six in 2000) received a long-form questionnaire that contained additional questions and provided more detailed socioeconomic information about the population, asking more detail than the shorter form that went to all respondents.

measure plumbing facilities have been a long-standing source of jokes and a major source of complaints.

There have also been complaints about the burden of housing-related questions. For example, many household respondents, particularly those who own their homes (about two-thirds of households nationwide) and those with a mortgage (more than two-thirds of homeowners nationwide), face a total set of about 30 housing questions. Other respondents complain that the time required to fill out the survey (estimated at 40 minutes) is too long.

The Census Bureau has responded to the concerns about ACS burden in a number of ways. In 2012, it asked the Committee on National Statistics (CNSTAT) to convene a workshop to consider the benefits and costs of the ACS for a wide variety of nonfederal users of the data products. That workshop considered both the burden of responding to the ACS questions and the importance of the ACS to the nation and the economy (National Research Council, 2013a).

Over the years, Census Bureau staff and outside organizations have carried out research on the costs and benefits of reducing the number of follow-up calls and visits (Zelenak and David, 2013). The results of that research led to a decision to implement some cutbacks. The bureau also established an ombudsman-type position (a "respondent advocate") to handle congressional and respondent concerns.

Since the 2013 CNSTAT workshop, the tempo of congressional interest has increased, particularly with regard to the mandatory nature of the survey and its burden on respondents. In response, the Census Bureau in 2013-2014 completed a review of the ACS content in terms of the need for each item. Based on that review, it recommended to the Office of Management and Budget that two questions be dropped: one on business or medical office on the property and one on the availability of a flush toilet (U.S. Department of Commerce, 2015). However, it recommended keeping the questions on hot and cold running water and bathtub or shower.

Congressional criticism has continued since that review. In his opening remarks at the workshop, Census Bureau Director John Thompson noted that at the 2016 Senate hearings on the U.S. Department of Commerce budget, a senator expressed concerns regarding the length of the ACS and asked why the Census Bureau could not get the necessary information from the private sector. At the same time, the House of Representatives passed appropriations bills in 2014 and 2015 that would have turned the ACS into a voluntary, instead of mandatory, survey.

STRATEGIES FOR REDUCING RESPONDENT BURDEN

The Census Bureau has conducted an active research, development, and evaluation program to address ACS response burden issues. The components of this program were summarized in a 2015 paper, *Agility in Action: A Snapshot of Enhancements to the American Community Survey* (U.S. Census Bureau, 2015a). The paper outlined a comprehensive and ambitious program to minimize burden for ACS respondents while still allowing the survey to respond to emerging issues by updating content as needed and maintaining high-quality data.

In her opening remarks at the workshop, CNSTAT Director Constance Citro stressed the importance of maintaining high response rates and relevant content and quality. Thompson expanded on this notion, stressing that the ACS "is extremely valuable to the country. It is used to allocate $400 billion of federal funds a year, and it is the only source of consistent data for many population groups, such as veterans." He also noted its importance to businesses: For example, at a White House event he attended, a number of technology companies reported on projects that combine ACS data with data for cities and other indicators. The companies had developed a rich array of applications, including one to create opportunities for disadvantaged individuals and identify schools that produced the right kind of skills and another to identify locations of affordable housing within reasonable distance of available jobs.

Steering Committee Cochair Joseph Salvo explained the committee decided to focus on four areas of investigation previously identified by the Census Bureau:

1. Building respondent support for the ACS through a communication and education strategy that focuses on respondents and considers stakeholder materials and efforts at marketing or branding the ACS with the goal of increasing participation by increasing understanding of how the ACS data are used.
2. Direct substitution of information from administrative records as a means of eliminating some questions.
3. Matrix sampling and other statistical methods that could reduce the number of individuals to whom the various questions are posed.
4. Changing the strategy for the collection of group quarters data.

This summary of the workshop is organized around a discussion of response burden and those four themes. Chapter 2 defines response burden and summarizes methods that have been employed and suggested to reduce the burden. Chapter 3 addresses means of improving response by building respondent support for the survey. The use of administrative records in addition to or as a substitute for the questionnaire is discussed in Chapter 4. Using matrix sampling and other statistical methods for reducing the number of respondents or the complexity of the questionnaire is the topic of Chapter 5. Improvements to the collection of information from group quarters are addressed in Chapter 6. The workshop's discussion of future directions for efforts to reduce burden is summarized in Chapter 7. The agenda for the 2-day Workshop on Respondent Burden in the American Community Survey is in Appendix A, and biographical sketches of the steering committee and presenters are in Appendix B.

2

Approaches to Reducing Response Burden

In his opening remarks, Steering Committee Cochair Joseph Salvo (New York City Department of City Planning) set forth a framework for the workshop that emphasized the importance of the American Community Survey (ACS) to the economy and the functioning of governments at the federal, state, and local levels. He also considered the threat to the survey posed by those who see it as an unnecessary burden and a threat to privacy. The importance of the survey, he said, "puts extraordinary pressure on the Census Bureau to not only educate the nation on the importance of the ACS and meeting the needs of our democracy, but to execute the survey in a manner that maximizes efficiency and minimizes burden."

Salvo emphasized that the steering committee adopted a broad view of burden. Burden is not simply the length of the questionnaire or the time needed to complete it, he explained, but also the perceptions of burden that come from many sources, including a respondent's views about government. He noted the perception of burden is difficult to measure, but it can be increased or alleviated by the materials that accompany the survey, the means of contact, and the perceived relevance or intrusiveness of the questionnaire. He challenged the participants to address the workshop goal of providing the Census Bureau with guidance for short- and medium-term solutions that do not require lengthy and/or expensive research.

CENSUS BUREAU CHANGES TO REDUCE RESPONDENT BURDEN

Deborah Stempowski (chief of the ACS at the time of the workshop)[1] observed that an all-inclusive definition or description of response burden does not exist—every person might identify the components of the definition differently. She noted that the estimate of 40 minutes to complete the survey is viewed by some respondents as intrusive, as is the number of contact attempts. Other aspects of burden include the ease or difficulty in providing answers to the questions and respondent concerns about the need for the information requested. Some respondents are concerned about perceived intrusiveness on the part of the government and question why the survey is mandatory, she said.

The report *Agility in Action: A Snapshot of Enhancements to the American Community Survey* (U.S. Census Bureau, 2015a) constitutes a plan for approaching the issues of response burden. Stempowski summarized a number of initiatives in the report that have formed the basis for the Census Bureau's approach: creation of the position of respondent advocate; fewer computer-assisted telephone interviewing (CATI) contact attempts; a new brochure, "Why We Ask"; refresher training for staff in contact centers and field representatives; some change in the content of the survey; reduction in the number of mail contacts; individual performance coaching for the field interviewers; and enhancement of the Internet part of the survey. She explained each change in more detail:

- **Respondent advocate** Stempowski reported that the position of respondent advocate, created in April 2013, was designed to resolve respondents' concerns. The advocate responds directly to issues raised by respondents and interacts with other stakeholders, including Congress. She said the advocate "is an ear to the ground" and helps guide improvements in data collection, the questions, and operations of the correspondence control unit and the call centers.
- **Fewer CATI contact attempts** The Census Bureau conducted research to model the possible effects of reducing the number of contact attempts. Stempowski noted that CATI software permits changing the parameters. After the research and *Agility in Action* report, revised contact stopping rules were implemented. The result was a reduction in CATI log-in hours of about 17 percent while the CATI response rate dropped only about 5 percent. The Census Bureau is now planning similar changes in its computer-assisted personal interviewing (CAPI) operations in June 2016, which will

---

[1] Deborah Stempowski became chief of the Decennial Management Division of the Census Bureau since the workshop and, at the time of this publication, had been replaced on an acting basis by Victoria Velkoff.

allow developing a score that reflects total contact attempts across mail, telephone, and personal visits.
- **"Why We Ask" brochure** A brochure has been prepared to provide respondents with information on why the ACS asks certain questions. The brochure addresses the benefits of responding in personal terms. Stempowski characterized the main message as "what is in it for me?"
- **Refresher training for staff in CATI contact centers and field representatives** The new training has been completed for staff in CATI contact centers and is scheduled to begin late in 2016 for field interviewing staff. The training reinforces the basic principles of reducing burden and treating respondents respectfully and professionally. It includes information on response conversion to persuade, without being too pushy, a reluctant person to participate.
- **Content changes** As noted above, a question on flush toilets has been deleted, as has the question on whether a business or medical office operates on the property. A question on computer and Internet use was revised to improve its currency.
- **Reduced mail contacts** The ACS employs a number of mail contacts in the process of soliciting responses. At the time of the 2015 National Research Council ACS review (National Research Council, 2015, p. 47), the first mailing was a prenotice postcard, followed by an advance letter that alerted sample members to the survey and encouraged participation. The letter was followed by a mail package, which included instructions for how to respond through the Internet. A reminder postcard was sent a few days after the mail package. Sample members who did not respond after the reminder postcard were sent a replacement mail package, which included a paper version of the questionnaire and a postage-paid envelope for a mail response. Instructions for responding by the Internet were also included. The package was followed by another postcard reminder. Sample members who did not have a telephone number that could be used for telephone follow-up received an additional postcard, alerting them that a field representative would be contacting them in person if they did not respond by mail or Internet. In 2015, the Census Bureau eliminated the prenotice postcard, and it has also accelerated the initial mailing date to increase the likelihood of self-response before the respondent receives the paper questionnaire.
- **Individual performance coaching for field interviewers** Though expensive, one-on-one coaching is valuable to provide feedback and reinforcing reminders, Stempowski said.
- **Internet instrument enhancements** In the past, the ACS had a single annual updating of the instrument at the beginning of the survey

year. A midyear updating each July makes improvements instead of waiting until the beginning of the next survey year. The Census Bureau is taking advantage of the midyear update in 2016 to (1) add security questions so respondents can create their own personal identification numbers (PINs), (2) highlight the write-in boxes to make them easier to see on a computer screen, (3) improve transitions through the instruments by consolidating and streamlining questions so they are easier to follow, and (4) add the "Why We Ask" text to the Internet instrument.

DEFINING, MEASURING, AND MITIGATING RESPONSE BURDEN

Scott Fricker (Bureau of Labor Statistics [BLS]) provided a brief overview about how response burden has been treated in the broader survey literature and shared the results of some of the burden-related research that he and his colleagues have been doing at BLS.

Response burden is important, he stated. In addition to ethical considerations of overburdening respondents, burden affects the quality of survey products. The continuing downward trend in response rates for most surveys has increased concern about the effect of burden on nonresponse, including panel attrition. He stated that establishment surveys are particularly concerned with delayed responses that affect initial estimates. Burden also can affect quality among survey participants, through item nonresponse and breakoffs, and it may cause less effortful, less accurate reporting by respondents.

Fricker said that dealing with the consequences of respondent burden has significant financial costs for survey organizations. Those costs include efforts to engage and secure the cooperation of sample members (through increased contact attempts or more elaborate and costly persuasion efforts), as well as procedures for dealing with suboptimal data (through editing and imputation). In his view, broader consequences include negative evaluations of surveys in general that negatively affect survey participation.

The importance of burden underscores the importance of appropriately conceptualizing and measuring it. Fricker described two approaches to measuring burden. The most common approach, historically, has been to equate burden with the length of the interview. This concept of "burden" relies on objective measures, such as the estimated total time, effort, and financial resources expended by the survey respondent to generate, maintain, retain, and provide survey data (U.S. Office of Management and Budget, 2006, p. 34); the interview duration (Groves and Couper, 1998, p. 251); or the number and size of the respondent's tasks (Hoogendoorn and Sikke, 1998, p. 189). Fricker said that in the absence of any additional information, these objective measures seem appropriate, especially survey length. However, the research is somewhat mixed in terms of how well they predict survey outcomes.

The second approach to measuring burden, he stated, is grounded in the psychological underpinnings of respondents' experiences in a survey. He referred to Bradburn's (1978) seminal work that identified four factors that contribute to respondent burden: (1) length of the interview, (2) amount of effort required by the respondent, (3) amount of stress experienced by the respondent, and (4) frequency with which the respondent is interviewed. In addition to underscoring the multidimensional nature of burden, Bradburn emphasized that "burdensomeness" is a subjective characteristic of the task, "the product of an interaction between the nature of the task and the way it is perceived by the respondent" (p. 36).

Along these lines, a number of researchers have focused on trying to capture the subjective nature of burden, most often using self-reports and interviewer notes. Self-reports measure respondent perceptions of a survey's characteristics (Sharp and Frankel, 1983; Hoogendoorn, 2004; Fricker et al., 2011, 2012); attitudes about the importance of the survey, about government, and similar entities (Sharp and Frankel, 1983); negative feelings (e.g., annoyance, frustration, or inconvenience) (Frankel, 1980); and perceptions of time associated with the response task (Giesen, 2012). Interviewers' notes are used to obtain respondents' complaints about survey burden (Martin et al., 2001).

These measures can be summarized in a conceptual model of burden. Fricker laid out the model used at BLS, which considers the objective burden and the characteristics of respondents (see Figure 2-1). The objective burden is measured by length, question content and layout, contact rules, collection mode, and persuasion strategies. The characteristics of respondents cover their cognitive capacity, motivation, attitudes about this survey or surveys in general, confidentiality concerns, and factors that might make the survey task more or less difficult, for example, household size.

Fricker described the BLS research focus on understanding respondents' subjective experiences of participating in the Consumer Expenditure Survey. BLS began by conducting a thorough review of the literature related to burden—both in the research on survey methods corpus and in psychological studies and studies of burden in medical caregivers—to identify areas likely to contribute to burden in the survey context. The study took the perspective that burden is a multidimensional construct, and it is important to identify features or dimensions that contribute to it. BLS then developed questions to assess those dimensions and administered them to respondents after their final interview (see Box 2-1).

With these data in hand, BLS evaluated the performance of burden-related items through five methods: (1) small- and large-scale analyses (cognitive and psychometric testing and field experiments), (2) multi-

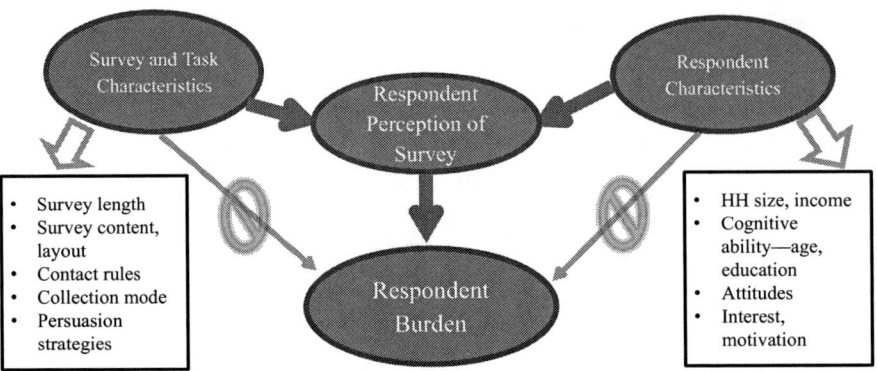

FIGURE 2-1 Conceptual model of burden.
NOTE: HH = household.
SOURCE: Scott Fricker presentation at the Workshop on Respondent Burden in the American Community Survey, March 8, 2016. Available: http://sites.national academies.org/cs/groups/dbassesite/documents/webpage/dbasse_173169.pdf [October 2016].

variate models of burden, (3) methods to produce a summary burden score, (4) associations with key survey outcomes, and (5) research into design features that impact burden dimensions.

Fricker summarized several findings. First, the research found consistent support for a multidimensional concept of burden—individual components or latent factors—such as respondent effort, item difficulty, and attitudes about length or sensitivity. All these factors contributed uniquely to respondents' overall assessments of survey burden. It also showed the importance of including respondents' subjective reactions in models: doing so improves model fit and strengths of association with survey outcomes. Looking at the individual contributions of objective measures of burden and more subjective metrics in these models, Fricker concluded burden is most highly associated with perceptual measures. Objective survey features, such as length and number of call attempts, have only a small direct impact on burden.

The major finding of the study, Fricker said, was that psychological experiences or reactions to those characteristics are the main drivers of burden perceptions. Fricker concluded the modes of data collection do not affect the overall structure of the model. The same factors have the same effect on burden regardless of whether respondents were contacted mostly by telephone or in person.

In terms of data quality, the BLS research found that intermittent participants who were contacted several times to solicit their response and

> **BOX 2-1**
> **Items Used by the Bureau of Labor Statistics to Assess Burden**
>
> **Perceived Burden**
> - How burdensome was this survey to you? (on a 5-point scale from "not at all" to "very")
>
> **Perceived Length**
> - Do you feel that the length of today's interview was too long, too short, or about right?
>
> **Perceived Effort, Interest, Sensitivity**
> - How difficult or easy was it for you to answer the questions in this survey?
> - How interesting was this survey?
> - How sensitive did you feel the questions I asked today were?
>
> **Perceived Frequency**
> - . . . number of calls you received—too many or a reasonable number?
> - . . . number of interviews—too many or a reasonable number?
>
> **Additional Items Tested**
> - Willingness to participate again; what if the interview was extended by 15 minutes?
> - Perceptions of confidentiality; usefulness of survey; time well spent, etc.
>
> SOURCE: Scott Fricker presentation at the Workshop on Respondent Burden in the American Community Survey, March 8, 2016. Available: http://sites.nationalacademies.org/cs/groups/dbassesite/documents/webpage/dbasse_173169.pdf [October 2016].

to accomplish a "refusal conversion" reported the highest burden. Interestingly, the data collected from these reluctant participants had little influence on the weighted estimates and the regression coefficients. In another test relevant to the ACS, BLS found using a split questionnaire, in which a subsample selected on a matrix basis received a questionnaire with fewer items, resulted in lower burden and in higher data quality.

Fricker turned to an evaluation of the ACS approach. He concluded that the approach is systematic, multipronged, transparent, and outcome oriented, and, therefore, it is likely to be productive. Appropriately, it focuses on the end result—whether the data are "fit for use." He applauded the selected hybrid approach, which considers both objective and subjective measures, as most likely to lead to additional insights and more targeted interventions. He said he supported the continued content review program.

Fricker suggested possible extensions to the ACS burden research agenda. For example, the Census Bureau might explore use of expert and interviewer ratings of items. He supported a project to solicit additional input from interviewers and added that assessment of the quality of interviewer observations and ratings—how well they track what respondents actually are thinking—might be productive. And he urged continued exploration of paradata.

In summary, Fricker stressed four points to consider when developing a future program to deal with response burden issues. First, perceptions of questionnaire length are affected by many factors, not length alone. Second, perceived length is a driver of burden, but there are many others. Third, the interaction of respondent characteristics with survey features should be a key consideration. Finally, when considering the way ahead, it is important to consider likely effects of intervention/design change on burden dimensions and how those design changes could be evaluated.

3

Improving Response by Building Respondent Support

This chapter summarizes the workshop sessions titled "Communicating with Respondents: Materials and the Sequencing of Those Materials" and "The American Community Survey: Communicating the Importance to the American People." The presenters addressed the Census Bureau's strategies for ameliorating respondent concerns by improving survey materials and providing compelling and accessible public information about the rationale for asking questions—a key component of the Census Bureau's research and implementation program to reduce respondent burden in the American Community Survey (ACS), as outlined in *Agility in Action: A Snapshot of Enhancements to the American Community Survey* (U.S. Census Bureau, 2015a). The Census Bureau communication strategy has mainly centered on improving the information on the ACS Internet instrument and on the ACS Website.

MAIL CONTACT STRATEGY AND RESEARCH

Elizabeth Poehler (Census Bureau) focused on the current ACS mailing strategy and recently conducted Census Bureau research on the topic.

Current ACS Strategy

In setting the stage for this discussion, Poehler outlined the sequence of contacts in the current ACS. The data for each monthly panel are collected over 3 months in a multimode sequential process. In the first month, she stated, the focus is on self-response, allowing either Internet or mail responses and continuing through the month. In the second month, the focus shifts to telephone interviews for those who have not responded. The third month focuses on in-person interviews with a sample of nonrespondents. Poehler discussed the self-response phase in her presentation.

The current ACS mail strategy begins with an initial package mailed to sampled addresses, inviting the potential respondents to respond via the Internet. In that package, the Census Bureau includes a letter, a Frequently Asked Questions (FAQ) brochure, an Internet instruction card, and a multilingual brochure. Approximately 7 days later, the respondents receive a reminder letter. About 14 days after that, addresses that have not yet responded receive a paper questionnaire package, which includes a letter, the FAQ brochure, and Internet instruction card, as well as the paper questionnaire, instruction guide, and a return envelope. (The Census Bureau plans to remove the instruction guide in spring 2016 from this mail package.) The process continues for nonrespondents. Four days after the paper questionnaire package, nonrespondents receive a reminder postcard. About 2 weeks after that, addresses for which the Census Bureau has a phone number receive a telephone call from the computer-assisted telephone interviewing (CATI) operation; those for whom no phone number has been identified receive an additional reminder postcard.

The process is involved and is traced with controls and metrics in order to maximize self-response, explained Poehler. One metric is to gather and assess the calls and written correspondence from those receiving the materials in order to gain insight into respondents' reactions to the mail materials. Though much of the correspondence asks questions to help understand the materials or clarify Internet access instructions, often the correspondence expresses concerns that tend to concentrate on (a) the legitimacy of the survey, (b) the intrusive nature of the questions, and (c) the perception of a negative tone of the materials, particularly questioning the mandatory language contained in many of the mail items sent to respondents. Poehler stated previous research has indicated that messaging about the mandatory nature of the ACS improves response rates, but some respondents bristle at the tone of the message. They express shock that the survey is required by law and feel threatened by the penalties and fines for failing to comply, she said.

The Census Bureau, in an attempt to address these respondent concerns while maintaining data quality at an efficient cost, has conducted research on ways to improve the mail materials and messaging to encourage self-response. This research has focused on addressing respondent burden, improving self-response rates through streamlined materials, and addressing respondent concerns about the prominent nature of mandatory messages on the mail materials. It has been based on laboratory testing with techniques such as focus groups and one-on-one interviews in order to solicit feedback on possible changes to the mail materials, such as better explaining the benefits of participating in the survey and modifying the look and feel of the materials. In connection with this research, expert feedback was also obtained.

Poehler stated that the research resulted in five high-level recommendations: (1) test visual design changes of the materials, (2) add deadline-related messaging on the envelopes, (3) eliminate the prenotice letter, (4) test additional mailings, and (5) tailor materials for non-English-speaking respondents.

Field Tests

Poehler described five field tests conducted in 2015: (1) Paper Questionnaire Package Test [March], (2) Mail Contact Strategy Modification Test [April]), (3) Envelope Mandatory Messaging Test [May], (4) Summer Mandatory Messaging Test [September], and (5) "Why We Ask" Insert Test [November]. The first two tests were implemented to field test suggestions from the messaging and mail package assessment research with the goal of increasing self-response by streamlining the mail materials, reducing the number of mailings, and cutting back on materials sent in those mailings. The next two tests focused on ways to soften mandatory messages by changing the visual design of the materials and explaining the benefits of participation. Finally, the "Why We Ask" Insert Test provided respondents with more information about why the ACS asked the questions it does. Poehler provided a high-level overview for each of the tests and their results.

Paper Questionnaire Package Test

The goal of this test was to reduce the complexity of the paper questionnaire package. The experimental design looked at modifying the mail package by varying (1) whether or not an instruction guide was included, (2) whether or not an insert that explained to respondents that they could choose to respond via mail or online was included, and (3) whether or not softened language that indicated the Internet as the preferred mode of response would increase Internet use. The test had four experimental treatments, each with 12,000 addresses and a control group with 238,000 addresses. This test found no significant differences between treatments and return rates, and no significant differences in item nonresponse rates, form completion rates, or response distribution. Costs were lower, but item nonresponse rates were nominally higher for treatments without the instruction guide. Based on this test, the Census Bureau recommended removing the instruction guide. There were smaller cost savings associated with removing the choice card and modifying the letter.

Mail Contact Strategy Modification Test

The goal of this test was to improve self-response rates by streamlining the mail materials. Streamlining consisted of three initiatives: (1) remove the prenotice letter and send the initial mailing earlier, (2) replace an initial reminder postcard with a letter that highlighted the respondent (user) identification, and (3) send the additional reminder postcard to additional addresses. The test design included a control group with 226,000 addresses and five treatment groups with a sample size of 12,000 each. The treatments varied three factors—whether or not a prenotice letter was included, whether or not the first reminder was a postcard or a letter, and to whom the additional postcard was sent. In the control version, the postcard was sent only to households without a phone number. In the experimental treatments, the card went to everyone, including households in the CATI operation. The results determined that eliminating the prenotice letter and sending the initial mailing earlier decreased total self-response return rates by 1.4 percentage points prior to sending the paper questionnaire mailing. However, prior to starting the CATI operation, there was no measurable difference in the self-response return rates. A reminder letter that highlighted the 10-digit user ID and included mandatory language increased total self-response return rates prior to starting CATI by 3.8 percentage points, while using a reminder letter but eliminating the prenotice letter and sending the initial mailing earlier increased total self-response return rates prior to CATI by a similar 3.5 percentage points. Finally, the tests found that sending an additional reminder postcard to addresses in CATI increased self-response return rates, but this was not translated into a noticeable change in the CATI response rates. Based on these findings, the prenotice letter was eliminated, the initial mailing was sent earlier, and the initial reminder postcard was changed to a letter, which began August 2015.

Envelope Mandatory Messaging Test

The goal of this test was to study the impact of removing mandatory messages from the envelopes. The envelopes in the initial mailing package and the paper questionnaire package contained mandatory messages. The control group received the envelopes with the messaging, and a test group received envelopes without the message. There were 24,000 addresses in each of the treatments. The results of this test pointed to the danger of removing the mandatory messaging. The test treatment had lower return rates by 5.4 percentage points prior to starting the CATI operation and, after all modes of data collection were complete, the test treatment had lower overall response rate of 0.7 percentage points. The costs of this action would be significant, Poehler said. Eliminating the mandatory messages from the envelopes alone is estimated to cost an additional $9.5 million if implemented due to the need to push response into more expensive modes.

Summer Mandatory Messaging Test

The goal of this test was to study the impact of removing or modifying the mandatory messages from a broader set of mail materials. Five treatments tested softening or removing the messaging. Each had a sample size of 12,000 addresses (see Table 3-1). The first two treatments were based on the current look and feel of the mail materials, and the control panel had basically no changes to the materials. Another "softened" control treatment tested removal of the reference to the mandatory nature of the survey in some places. The final three treatments had visual design changes and an expanded presentation of the benefits and uses of the ACS based on the messaging and mail package assessment research and consultation with external experts. The first "revised design" had a redesigned look and feel

TABLE 3-1 Summer Mandatory Messaging Test Treatments

Test Treatment	Strategy
Control	• No change in materials
Softened Control	• Mandatory messages removed from initial letter, mail package letter, postcards, and envelopes • Mandatory messages kept in FAQ brochure, reminder letter, instruction guide
Revised Design	• Redesigned envelopes, use of bold lettering, highlighted boxes, "Open Immediately" • Strong mandatory language
Softened Revised Design	• Revised design used • Mandatory messages removed from postcards and envelopes • Mandatory messages softened in letters (plain text)
Minimal Revised Design	• Revised design used • Mandatory messages removed in all materials except initial letter • Mandatory messages in initial letter on back of page, in small font

SOURCE: Elizabeth Poehler presentation at the Workshop on Respondent Burden in the American Community Survey, March 8, 2016. Available: http://sites.nationalacademies.org/cs/groups/dbassesite/documents/webpage/dbasse_173178.pdf [September 2016].

of the materials but included strong mandatory language. The "softened revised design" included the revised design but removed messaging related to the mandatory messaging from postcards and envelopes and softened it in other places. The "minimal revised design" removed the mandatory messages in all of the materials except the initial letter. In the initial letter, the mandatory language was on the back of the letter in small font. Poehler showed an example of the envelope change. The control initial envelope (see Figure 3-1) has bold language saying the response is required by law.

The redesigned initial envelope (see Figure 3-2) added the Census Bureau logo and a message to open immediately on the right-hand side. The revised design states, "Your response is required by law." In the softened revised design and the minimal revised design, that language was replaced with "your response is important to your community."

Poehler also presented an example of the current and redesigned control initial letter. The revised design letter adds the Census Bureau logo, uses bulleted lists, enhances the Website address by putting a box around it, uses bold lettering, and adds text to appeal to the respondent's sense of community. The softened revised design letter is nearly identical to the revised design except the sentence about the mandatory nature of the survey is not

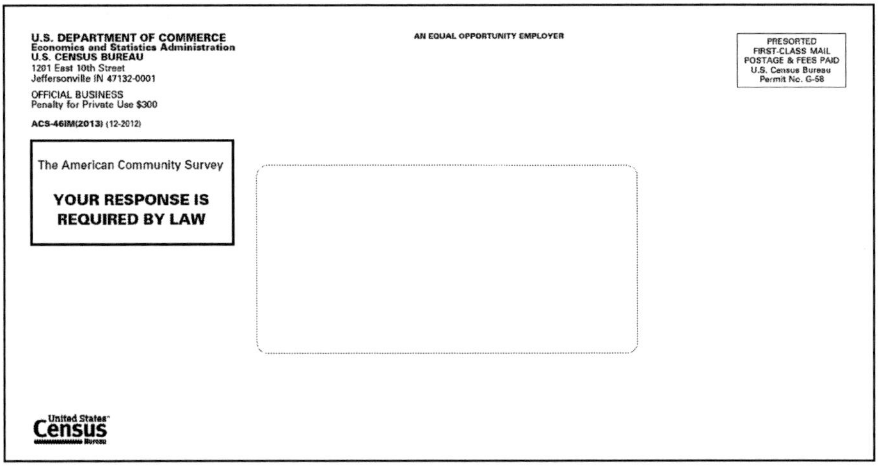

FIGURE 3-1 Control initial envelope with bold mandatory messaging.
SOURCE: Elizabeth Poehler presentation at the Workshop on Respondent Burden in the American Community Survey, March 8, 2016. Available: http://sites.nationalacademies.org/cs/groups/dbassesite/documents/webpage/dbasse_173178.pdf [September 2016].

IMPROVING RESPONSE BY BUILDING RESPONDENT SUPPORT

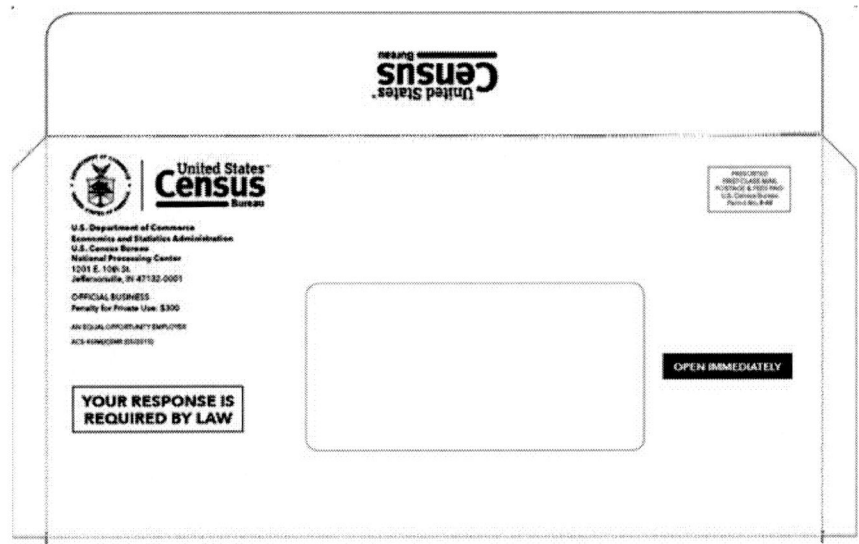

FIGURE 3-2 Redesigned initial envelope. and messaging.
SOURCE: Elizabeth Poehler presentation at the Workshop on Respondent Burden in the American Community Survey, March 8, 2016. Available: http://sites.nationalacademies.org/cs/groups/dbassesite/documents/webpage/dbasse_173178.pdf [September 2016].

in boldface type. Another version removes the mandatory sentence. Overall, the results from this test indicated that reducing the frequency and visibility of mandatory messages reduces response rates.

"Why We Ask" Insert Test

The goal of this test was to study the impact of including a flyer in the paper questionnaire mailing package explaining why questions are asked in the ACS. The control treatment did not include the flyer. One experimental treatment included the flyer, and a second experimental treatment included the flyer and removed the instruction guide. Each of these treatments had 24,000 addresses. The results from this test were expected to be available in June 2016.

Future Research

Poehler reported on two research initiatives under consideration. First, the Census Bureau is looking into testing the use of targeted digital advertising to deliver video and static-image advertisements to sampled addresses. The advertisements would be intended to create positive associations with the Census Bureau's work generally and the importance of completing a survey. They would not directly link to or mention the American Community Survey. A second initiative would explore findings from social and behavioral sciences to incorporate into the mail materials and the messages used to encourage people to self-respond.

IMPROVING RESPONSE TO THE AMERICAN COMMUNITY SURVEY

Donald Dillman (Washington State University) congratulated the Census Bureau on its experiments this year. However, he said, more could be done to improve ACS self-administered response.

As background, Dillman reported on a series of tests in the 1990s to assess 16 factors in an effort to improve mail-back response rates to decennial census forms (Dillman, 2000, pp. 298-313). He reported only five of them improved response: (1) respondent-friendly visual design, (2) prenotice letters, (3) postcard thank-you reminders, (4) replacement questionnaires, and (5) prominent disclosure on the envelope ("U.S. Census Form Enclosed: Your response is required by law"). The first four items have been shown to make a difference in all mail-back surveys, he noted, while the "required by law" effect was census specific (and came from business survey research). Moreover, other factors, including multiple contacts, produced an initial 58 percent response. The mandatory response notice added only modestly (9 percentage points) to this amount in noncensus year tests.

Dillman reported on a 1991 survey on nonresponse to the 1990 decennial census. People gave one of five reasons: (1) some did not remember receiving the form, (2) some received it but did not open it, (3) others opened it but did not start to fill it out, (4) still others started to fill it out but did not finish, and (5) a few completed the form but did not send it back. According to Dillman, these findings justify the need for multiple contacts, which are very powerful in boosting response by getting people to start and/or finish responding.

Dillman asserted Census Bureau survey sponsorship is a positive factor for response, "probably the most desirable sponsorship of any that one could have for getting response from the general public," because of its credibility compared with other organizations. Despite this positive aspect, obtaining response in a Web-push methodology (where a Web response is requested with an offer of mail later in order to recruit different demographics) is much more difficult than getting responses to only a mail-back procedure. He summarized 10 university-sponsored tests in multiple states that produced mean response rates of 43 percent for a Web-push methodology versus 53 percent for a mail-out, mail-back approach (Dillman et al., 2014). He explained the reason for lower response is that switching from one medium of communication (mail contact) to Web response requires special effort on the part of the person who received the request.

He illustrated the effect of three methods with several of his experiments over the course of about 2007 to 2012 (see Figure 3-3). In every case, response rates were higher with mail only. Although he agreed that the Web-push method is needed, he noted that these data illustrate the difficulties of the task.

Dillman then discussed how to make a communication sequence effective. He concluded it is necessary to design all visible aspects of mail contacts—the outside appearance of the envelope or card (size, shape, print), the message (letter) requesting a response, enclosures, the Census form cover pages, and the actual questions—in mutually supportive ways.

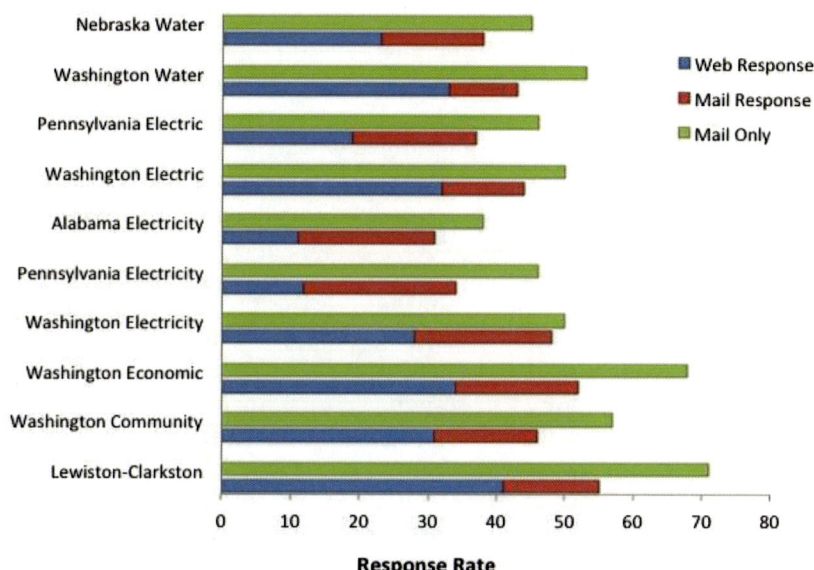

FIGURE 3-3 Comparison of response rates across 10 studies.
SOURCE: Don Dillman presentation at the Workshop on Respondent Burden in the American Community Survey, March 8, 2016. Available: http://sites.nationalacademies.org/cs/groups/dbassesite/documents/webpage/dbasse_173168.pdf [September 2016].

He listed factors that work against individual effectiveness of contacts. These factors include keeping the same outside appearance on most mailings, repeating the same content so new information and appeals cannot be added, including enclosures not relevant to most people who will respond, failing to convey the importance of each household's response, and failing to use new opportunities/places for effective persuasion in later contacts. In summary, he stated the goal is to avoid sameness of arguments and not to let each contact become unfocused with too many disparate or repetitive enclosures.

In order to implement these improvements, Dillman offered suggestions to modify the five ACS contacts outlined in Poehler's presentation to avoid sameness of arguments and letting each contact become unfocused. He pointed out that currently there are five contacts for obtaining responses to the Web version plus a mail option.

- **First contact** In the first contact, he would remove elements that interfere with the focus and add wording about how a person's response helps America. He would modify the envelope to refer to a "U.S. Census Form" rather than "The American Community Survey," he said, because the U.S. Census form, unlike the ACS, is an established brand in most Americans' minds. He would simplify the mailing by removing items. He said he would remove the very general "Frequently Asked Questions" document and place the content into the letter, perhaps on the back page. He would also take out the multilanguage brochure, which, in error, says the Census form will arrive in a few days. He said he would replace these items with the "Why We Ask" brochure, which gives concrete examples of why the ACS is important. Finally, he would change the cover letter so it no longer appears mass produced, taking out the salutation and the message from the director. He would date the letter, because all culturally targeted communications have dates on letters sent sequentially. Additionally, he suggested the letter explain why people are required to respond, inform them the response applies to everyone living at the address, and tie the letter to the "Why We Ask" brochure.
- **Second contact** Dillman stated he was pleased with the Census Bureau's changes in the second contact, with a new letter replacing a prenotice letter and reminder postcard. The reminder postcard was a carryover from the mail-only request. It could not provide the name of survey and login information, he said, so a respondent was pushed back to the first mail-out, thus increasing the burden of figuring out how to respond online. He observed that this change illustrates how Web-push methods need different contacts than

mail-only approaches. It resulted in an Internet response improvement of about 5 percent, with total Internet and mail response increasing by about 3.5 percent.

- **Third contact** He suggested major changes in the third contact which, as detailed in Pochler's presentation, consists of seven pieces of paper: outgoing envelope, frequently asked questions, 16-page instruction booklet, login card, message from the director, paper questionnaire, and return envelope. He concurred with the results of the Census experiment to remove the instruction book and the choice card, shifting the card content to the letter. He would again add the "Why We Ask" brochure. He suggested changes to the first page of the ACS questionnaire, which he stated is not conveying to people the reasons that they should respond. He would add the following across the top: "The American Community Survey—Producing quality-of-life statistics that communities in every state depend on to assess their past and plan for the future." He would also insert material from the "Why We Ask" brochure to explain what the ACS is, with this content aimed at explaining the ACS to a second or third person in the household who may become involved at this stage of responding.
- **Fourth and fifth contacts** Dillman said he would leave the fourth contact as a postcard format, but leave out the emphasis that a response is required by law and that an enumerator will visit. He would change the fifth contact to a letter that focuses on why response is required and that states it is the last letter prior to telephone or an in-person visit.

ACS RESPONDENT MATERIALS AND SEQUENCING: APPLICATION OF A RESPONSIVE AND ADAPTIVE SURVEY DESIGN FRAMEWORK

Andy Peytchev (University of Michigan) introduced his remarks by stating that the relationship between survey design and survey burden is complex. On the one hand, response burden is not a well-defined concept. It may have different dimensions, it is difficult to measure, and it depends a lot on the circumstances of the respondent. On the other, design is important to the challenge of ameliorating burden in that it can affect burden by changing mode, asking fewer questions, asking the questions better, and using matrix sampling, among other things. Good survey design needs to consider simultaneously costs, burden, quality of the resulting estimates, and intended use for the estimates. Thus, the relationship is extremely complicated, he said, which means the best design may not be known and may never be achievable. Still, it is important to strive for a better design, and in the process develop solutions that may improve the survey and get closer to the survey objective. In the end, however, there may not be one specific design prescription, for multiple reasons.

Depending on the objectives and their relative importance, the best design will depend on several factors, Peytchev said. First, it is necessary to define how important burden is relative to the other factors, which may vary by question. The multiple objectives that should be optimized are straightforward—avoid undue burden, reduce cost, and increase response rates. The solutions are not clear, given the near-infinite pool of design options, combinations, and permutations.

In a survey like the ACS, design is an overwhelming challenge, he observed. Many features and constraints on the ACS limit what can be fixed within the design, including the multiple modes, the mode sequence, and the data collection periods. Those limits become even more evident when both Web and mail responses must be collected within the same 1-month period.

The reaction of respondents to the design varies and, like the perception of burden, may change over time. The burden of a mail questionnaire as perceived 10 years ago may be different now, especially with the Web-mode option. Thus, burden may need to be constantly reevaluated. Peytchev pointed out that perception of burden may also vary across sample members. What one person may see as burdensome, another person may see as motivating.

He concluded that the complex interrelationship between design and burden, and the added complexity of the ACS, challenge the usual approach to survey redesign. The usual approach is to try to identify possible main factors to change, package them into several changes at the same time, and mount a standalone experiment to test them. Sample size determines how many features can be disentangled and how many interactions can be identified within the experiment.

The ACS experiments in 2015, he observed, were a bold departure from the usual approach and definitely in the right direction. The 2015 approach should be part of a permanent design, he said. The key was the responsive design framework, developed by Groves and Heeringa (2006). He suggested modification of the framework for purposes of the ACS by inserting consideration of burden and ACS features into each of the framework's four steps: (1) pre-identify a set of design features potentially affecting burden, costs, and errors of survey estimates; (2) identify a set of indicators of the burden, develop cost and error properties of those features, and monitor those indicators in initial phases of data collection; (3) alter the features of the survey in subsequent phases (and monthly sample releases) based on cost–error tradeoff decision rules; and (4) combine data from the separate design phases (monthly sample releases) into a single estimator. By stating the framework in this manner, it is possible to consider burden within the key objectives and the responsive design. While cost and survey errors may be prominent feature outcomes within this design, Peytchev suggested inserting burden into what to optimize for, including indicators for burden; altering features that may affect burden; and then combining the data from these multiple phases.

The framework can be very relevant for the ACS, he said, in the following sequence: view the ACS as a 1- and 5-year survey, with multiple/monthly sample releases; learn from one sample release to the next; leverage the continuous use of tests to measure the impact of individual features; employ as a permanent feature of the survey; and ensure that survey errors are evaluated, in addition to cost, burden, and other outcomes. In this manner, over time, it is possible to converge to a more optimal design.

Another key feature of the responsive design framework is the focus on an evaluation of the survey errors, not just the burden. Evaluating the error properties of the survey estimates may result in a limited concern for nonresponse because, within subgroups, there may be greater concern for measurement error. In accord with the framework, Peytchev said that the 2015 Census Bureau studies referred to by Poehler actually evaluated the effect on survey estimates when they omitted the instructions.

Adaptive survey design is a related notion, Peytchev observed. A key feature of adaptive survey design is to acknowledge heterogeneity—people are different, and each person or group of people may need different treatment over time during the data collection period. In this environment, there may be variability in how they perceive burden and how they respond to different design features within the content materials. These variable responses would be considered in tailoring the designs at the sample address or the subgroup level. The materials, content, and sequence of these materials would be taken into account in the resulting design.

He challenged the Census Bureau to adopt the theoretical framework and a responsive and adaptive design approach and to implement the approach over time consistently by testing different factors. The theoretical framework would be based on Leverage-Salience Theory (Groves et al., 2000), which postulates that different people participate or do not participate for different reasons. This theory has relevance to burden on the ACS in that some motivating factors may reduce perceived burden.

Another standard is the Compliance Principles (Cialdini, 1988), which postulate several means of assuring compliance (response). These means include *authority* (e.g., different ways to invoke the government's involvement and mandatory nature); *reciprocation* (e.g., "you have benefited from the services that result from the ACS"); *consistency* (e.g., "as a good resident in your community..."); *social validation* (e.g., "98% of selected households complete the ACS"); *scarcity* (e.g., "your address has been selected to represent many others in your community"); and *liking* (e.g., start the introduction with something positive).

In summary, Peytchev observed the ACS is already one of the most innovative survey designs among the major federal surveys in that it has production sample replicates for experimentation, a formal multiphase design with multiple modes, and double sampling of nonresponse. He stated these complexities add value to the survey, and they also make it more flexible. Considering a more formal adoption of the responsive design framework to continuously improve the survey may be of benefit, and, he said, the ACS is uniquely set up to leverage these capabilities and exploit the flexibility.

COMMUNICATING WITH RESPONDENTS: MATERIAL AND SEQUENCING IN THE ACS

Nancy Mathiowetz (University of Wisconsin–Milwaukee [emerita]) focused on two questions related to the documents in the first mailing: to what extent is the information within and across each of these documents consistent, and is the most important information clearly conveyed.

Mathiowetz first discussed the mailing envelope used in the 2016 production. She questioned whether all seven distinct pieces of information on the envelope are needed. For example, both the U.S. Department of Commerce and the Census Bureau are identified. The envelope indicates the Census Bureau is an equal opportunity employer. She acknowledged eliminating pieces of information may be very difficult, but currently it is difficult to know which piece of information is the most important. She stated that the redesigned envelope has much better branding and better identification of the Census Bureau.

She then assessed the letter. At first glance, she said, the respondent sees fairly dense prose and off-putting language. The term "randomly" appears twice. While the first definition of the term in Webster's dictionary is "proceeding, made, or occurring without definite aim, reason, or pattern," the meaning of "randomly" for purposes of the ACS is that the person was selected with purpose by a scientific process, she noted. She urged the Census Bureau to consider how the lay population interprets terms like this.

The letter, she assessed, contains contradictory information. It portrays a sense of urgency when, in order to push people toward the Web, it says to complete the survey online as soon as possible. It also says the agency will send a paper questionnaire in a few weeks—mitigating the sense of urgency in the first part of the text and sending a mixed message. Another point of confusion is that the household is selected for the sample, but a person is mandated by law to respond. She asked how the Census Bureau could convey the idea of a mandate to respond when an individual is not the respondent.

She opined that the new letter is visually more appealing and brands the Census Bureau. One problem is that the return address on the letter—Washington, D.C.—does not match the return address on the revised envelope, which says Jeffersonville. Other improvements to the new letter include dropping the issue of timing and no longer pushing the respondent to the Web and then announcing a paper questionnaire in a few weeks.

Agreeing with Dillman, she pointed out an inconsistency in the current brochure. The first part says the person will receive an American Community Survey within a few days. This should not be included in a mailing that tries to push people toward a Web survey. In her view, it would be better to eliminate this brochure to have a consistent message.

COMMUNICATING THE AMERICAN COMMUNITY SURVEY'S VALUE TO RESPONDENTS

Andrew Reamer (George Washington University) presented several ways of raising the perception of the value of the ACS as part of a campaign to reduce burden. Along with previous presenters, he questioned the notion of burden, which he stated has both technical and political meanings. In political terms, burden suggests that the respondent is a victim. The term is used in a political sense by people who are unhappy with the ACS and believe it is intrusive. He encouraged alternative wording when possible.

By way of background, the final report of the congressionally established Commission on Federal Paperwork (Commission on Federal Paperwork, 1977) surfaced the notion of burden. The commission concluded that a wall of paperwork had been erected between the government and the people and that countless reporting and record-keeping requirements and other heavy-handed investigation and monitoring schemes had been instituted based on a faulty premise that people will not obey laws and rules unless they are checked, monitored, and rechecked. In this context, the commission defined different types of paperwork burdens. Economic burdens include the dollar cost of filling out a report as well as the costs of record-keeping systems. Psychological burdens include frustration, anger, and confusion.

Reamer offered ideas to raise the perception of value of the ACS. His suggestions included revising the tagline; including an overarching framework of uses; noting the ACS origins; emphasizing the notion of the community getting its "fair share"; indicating the community response rate; redesigning the "Why We Ask" material to broaden the scope and highlight examples; and testing use of the Census Partnership Program with the ACS.

To Reamer, the tagline—"How your responses help America"—is too general. It should focus on how responses help the community, state, and nation to give a sense that it is about the respondent's community as well as the nation. Similarly, the framework of uses should reflect the ACS role in improving the economy, ensuring efficient and effective government, and sustaining democracy. He encouraged communicating these messages to the respondents.

With regard to efficient and effective government, Reamer suggested making clear that the idea of the ACS originated with James Madison. He proposed indicating that the ACS is the current iteration 226 years later of what Madison asked in 1790: to add questions to the decennial census so that Congress might legislate on the basis of the circumstances of the community. Reamer further emphasized the notion of a community getting its fair share of private sector goods and services, jobs, and government grants and government services. In line with the current emphasis on behavioral economics, he encouraged the Census Bureau to consider putting the response rate on communications to establish a social norm around response.

He critiqued the "Why We Ask" publication and urged its redesign. He said the document presents good examples but buries them in the text and makes no mention of the uses of the ACS for democracy (e.g., the ACS citizenship question is used to draw congressional district boundaries, and Section 203 of the Voting Rights Act mandates the use of the ACS language questions). He suggested highlighting the examples and adding an example about legislative boundaries.

Lastly, Reamer suggested a test in which the Census Bureau recreates its decennial census partnership in a few communities in the near future in order to let respondents know they can call someone in their neighborhood to affirm the legitimacy of the ACS and to explain the value of filling out the questionnaire. He observed that this would be a relatively low-cost initiative.

DISCUSSION

In response to an invitation from Linda Gage, cochair of the workshop steering committee, members of the audience offered questions and comments on the communications topic. One commenter noted the ACS letter now prominently features the fact that the ACS data determine the distribution of some $400 billion; however, given the current climate, some might think this is wasteful spending. Use of the figure might do more harm than good for some segments of the population. In response, Poehler clarified the figure appears only at the bottom of the experimental letter and in the letter currently in use. The language is still to be evaluated. Reamer further responded the money will be allocated one way or another and that participating in the ACS ensures a person's own community gets its fair share.

Another participant stated the success of the stimuli to respond quickly and with the cheapest mode depends on the timing of these iterations of mailings. Research in the context of the census program has shown that timing matters.

A participant noted presenters discussed proposed measures to downplay the mandatory nature of the survey. The person questioned whether it is a disservice to avoid informing people that they may get a fine for not responding. Poehler said language in at least one place on all of the tested mail materials tells the respondent that the ACS is required by law and the potential fine. The tests varied how much the Census Bureau emphasized or deemphasized that messaging, but it is required by law that a respondent is told if a survey is either mandatory or voluntary. Reamer added the amount of the potential fine—up to $5,000—is also a consideration. He stated that congressional opponents of the ACS always bring up the $5,000 fine, but the Census law passed in 1976, and still on the books, limits the fine to $100 for refusal to fill out the survey and $500 for false responses. The 1976 law was overwritten in the 1980s by comprehensive crime control legislation that pushed the fine up to $5,000. He contended that the $5,000 fine contributes to the perception of burden.

A participant asked about the relationship of respondent burden to the device on which respondents respond to the survey, inquiring if the ACS Internet instrument is optimized for completion on a mobile device. Lower-income individuals are more likely to access the Internet via a mobile device rather than through a desktop or laptop computer and a high-speed Internet connection at home. Whether or not the respondent can respond on the individual's device should be taken into account in addressing burden, according to the participant. Poehler responded that the Census Bureau conducted studies about people using smaller devices to access the Internet about 6 months ago. The study found problems in that respondents were pinching and zooming multiple times for every screen. Based on these findings, the Census Bureau redesigned and optimized the ACS for smaller devices. That redesign has been implemented.

A participant stated part of the burden is invasiveness, of which the telephone and the in-person interviews are a big component. The participant asked if the Census Bureau knows whether the complaints are about these types of interviews and whether it would make sense to go straight to the computer-assisted personal interview or extend the mail collection. Are there are other methods for self-response that seem less invasive? Poehler answered that the Census Bureau recently studied and addressed the burden associated with telephone calls and in-person visits, curtailed the number of attempts made by phone, and stopped visits when a threshold related to perceived burden has been reached. The Census Bureau has not thoroughly investigated whether to expand mailing beyond the current practice because of the telephone and visit burden.

Mathiowetz suggested one-size-fits-all is perhaps not the best approach. Consideration should be given to Census tracts and zip codes in an adaptive design framework. This would mean, for instance, that people with no Internet access would not be invited to respond on the Internet. It would call for modifying the sequencing, perhaps removing the telephone contact because it is seen as a fairly intrusive form of communication. She said that she is not aware of current research at the Census Bureau looking at these issues more microscopically. Dillman noted telephone contacts amount to about 7 percent of the total, which is not a very large amount. It may be possible to cut down on frustration by pushing the Webmail a little more and going straight to in-person interviewing.

A participant asked for clarification about the size of the problem. How many people are calling and complaining? What percentage of respondents are complaining? What are the gains in reduction and burden from some of the changes that have been introduced? A participant from the floor responded that the number of complaints is small; the problem is not the magnitude but the party doing the complaining. When the complaint comes from the congressional district of the chairperson of the budget committee, for example, the importance of the perception of burden is larger than the small numbers would suggest. Poehler added that, from a pure metrics perspective, because the volume is small, the effect is very difficult to measure. When the Census Bureau implemented the recent changes in treatment, it received about three more phone calls per month than in previous months. It was not possible to attribute this increase to the change.

Reamer commented on the tension between research findings that show an increased response rate when the mandatory response is emphasized and the program to test softening the mandatory message. He said the Census Bureau had to test softening the message for political reasons.

Another participant, a member of the Census Scientific Advisory Committee, emphasized the minuscule number of complaints. She further pointed out that her understanding is that no one has been fined for not responding to the ACS.

A participant who had been a respondent for the ACS expressed pleasure that the first postcard is now no longer a part of the mail-out because she found it confusing. Although glad to be able to respond online, she said she wished that she had more preparation for some of the questions. The nature of some of the questions—such as income in the past 12 months—cannot be answered from taxes from the previous year but only by aggregating paystubs. The need to obtain information to answer the questions meant leaving the computer. In her view, it would have been helpful to have the paper form to view the questions ahead of time.

COMMUNICATING THE IMPORTANCE OF THE ACS TO THE PUBLIC

In introducing this panel session, Nancy Mathiowetz stated that the workshop steering committee wanted to address any gaps in the Census Bureau's research. In terms of how the ACS is communicated to the U.S. population, there are questions concerning what stakeholders and the public in general know about this survey, what steps the Census Bureau can take to increase its visibility, and how the bureau can develop a long-term incremental marketing strategy to brand the ACS. She noted the discussants in this session represented the commercial marketing sector and market research organizations.

BRANDING TECHNIQUES

Sandra Bauman (Bauman Research & Consulting) noted branding is most often associated with the commercial world and the roots of branding are associated with commercial goods. Other types of organizations have used branding and capitalized on its tenets to make their communications more effective. She said branding is germane to the ACS as a tool to ameliorate respondent burden.

Bauman cited several instances, including the presidential campaign of Ronald Reagan, when political candidates and parties have used branding. Nonprofits and causes have used branding and marketing as well. Personal branding, such as a person's LinkedIn profile, is also employed. She emphasized a brand is much more than a logo or a graphic identity. She referred to advertising expert Walter Landor, who famously called a brand a promise in that it identifies and authenticates a product or service, and, in so doing, delivers a pledge of satisfaction and quality. Bauman defined a brand as, first, a promise that when delivered results in positive feelings of satisfaction and, second, a collection of perceptions in the mind of the consumer. It is different from a product or service in that it is intangible and is a psychological construct—the sum total of everything that the audience or customer knows about, thinks about, and feels about the brand.

Branding is related in several ways to the ACS, she said. A branded ACS can have greater impact than a nonbranded one. Branding the ACS is a vital part of spreading awareness about the survey. It can address barriers and obstacles to participation and serve as a bridge between the Census Bureau and the community.

She summarized five building blocks that make an effective brand:

1. Develop and be able to articulate a positioning. A positioning considers the attributes of the product or service and the benefits that customers can get from it. For the ACS, this building block means finding a benefit important and personally relevant to respondents and determining how to leverage it. According to designer David Galullo (2013), the positioning may be about the connection the seller (in this case, the Census Bureau) has with its employees and customers. Both want to feel like they are part of something larger.
2. Tell a story, bring it to life, and link it to personal value. It has to be believable, relevant, unique, and motivating. The story must persuade with reason but motivate with emotion.
3. Elevate the benefits of the brand by going from the functional to the emotional.
4. Deliver on the brand promise with every experience—authenticity and transparency build trust.
5. Make every touch point consistent in tone, language, look, and feel.

Bauman discussed how ACS participation could be reframed to overcome obstacles in the current context that include government distrust, concerns about privacy, and lack of awareness of the survey. She recommended that the Census Bureau assess the branding campaign and communication strategy employed in the 2010 decennial census program. Also, she said, the bureau should gather and publicize people's stories about how they and their communities benefited from the ACS, which can help to humanize the survey.

Bauman also emphasized the importance of building pride in being chosen for the survey and having participated. She suggested emphasizing exclusivity and how it is a privilege to be chosen to participate in the survey. She pointed out that people take pride in giving blood or voting and are motivated by stickers that announce their action to the world.

In summary, Bauman said, it is important to make respondents feel like they are part of something larger than themselves.

MESSAGING

George Terhanian (NPD Group) focused on the science of messaging and the progression in marketing from messaging about the features, to a focus on the benefits, to messages that evoke emotions, to messages that illustrate core values. He critiqued the current messaging found in materials provided to ACS respondents and, drawing lessons from consumer research and studies of business' approach to branding, commented on several statements in the materials.

Message: *The ACS is sent monthly to a small percentage of the population, with approximately 3.5 million households per year being included in the survey.*

Terhanian's comment: The message should be respondents have been selected at random, but they are special and part of an exclusive group. The fact that respondents are in the survey by invitation only is a message. He likened a message stressing exclusivity to the message used by Google when it first recruited Gmail users and how Nielsen, via the Nielsen Ratings, positioned its service for decades. He stated that respondents should be made to feel as though they have won the lottery. Certainly, the challenge is to communicate the benefits of the ACS (and participation in it) in a way that resonates at a personal level. He suggested that reducing interview length to accommodate respondents reinforces the message that they count and shows respect for their time and effort.

Message: *However, the entire country benefits from the wealth of information provided from this survey of over 11 billion estimates each year for more than 40 topics covering social, demographic, housing, and economic variables.*

Terhanian's comment: He challenged using the word "however." He suggested the positive message is that the entire country benefits through the participation of the 3.5 million lottery winners, the lucky ones others trust, depend on, and even envy. Their individual participation leads to direct benefits for themselves, their communities, and their country.

Message: *The data that the ACS collects are critical for communities nationwide—it is the only source of many of these topics for rural areas and small populations.*

Terhanian's comment: He suggested specifying the benefits to the individuals within the communities and, more specifically, to the 3.5 million respondents. The current messaging seems to suggest participation in the ACS is a necessary evil.

Message: *. . . Target, JC Penney, Best Buy, General Motors, Google, and Walgreens use ACS data for everything from marketing to choosing franchise locations to deciding what products to put on store shelves. Because ACS data are available free of charge to the entire business community, the program helps lower barriers for new business and promotes economic growth.*

Terhanian's comment: He said this is a wrong message because respondents do not relate to its "great for businesses" tone. They are interested in benefits to individuals and whether the information creates more jobs for working people. The message should be how the businesses are using the data to benefit consumers. For example, do they use the information to ensure they have in stock the products people need and want, which can vary by region? The ACS should communicate individual (consumer) benefit.

Message: *First responders and law enforcement agencies use ACS data.*
Terhanian's comment: Although this fact is very important and brings associations with family, safety, and peace of mind, Terhanian commented that disasters and emergencies are rare events. Instead, the message should focus on the close-in benefits to individuals.

Message: *Your response to the survey is required by law.*
Terhanian's comment: The message should be softer without diluting the core message that is profoundly important.

He summarized by saying that the ACS's main message should be that people count and that is why they should participate.

CONNECTING WITH PEOPLE AND COMMUNITIES

Betty Lo (Nielsen) gave examples of campaigns and the communications that Nielsen has leveraged to connect with consumers and communities through the company's Community Alliances and Consumer Engagement team. The team creates relevant messaging that the company can use to connect with those consumers.

Nielsen focuses on multicultural communities mainly because it has been found—through various advertising and brand awareness studies—that these communities are less likely to be aware of Nielsen and to participate in a Nielsen survey or a study. The methodology is to create a message that resonates with all audiences, and then put a cultural nuance on the message to connect specifically with diverse communities. In this way, Lo said, Nielsen's challenge is similar to the objective of changing the paradigm around how people think of the census and the ACS from perceiving a burden to understanding that responding helps them.

Nielsen started by developing a market-by-market strategy. This process was a challenge because Nielsen is prohibited from advertising on the media that the company measures. For example, the company cannot use television advertising, but with the approval of the Media Ratings Council, received a reprieve for an extended year to advertise on radio. Radio has been one of the most effective ways in which to connect with the community, Lo said.

As part of its decision-making process for selecting markets and communities, Nielsen annually conducts a brand awareness survey that measures how consumers perceive the Nielsen brand. The survey probes familiarity with the brand and the likelihood of participating on a panel if asked. The team also considers recruitment results, estimates of the universe from census data, and business partners. Lo speculated that, if adopted, this approach could bolster ACS communication and the ACS advertising campaign.

Having selected the communities and the approach, Nielsen seeks to leverage the community by engaging thought leaders and using earned media to get the message out into the community so that it resonates and connects and so that consumers take ownership. The goal of this approach, Lo explained, is to create a conversation where other people are talking about Nielsen rather than Nielsen having to talk about themselves in a typical direct marketing or advertising way.

The process has resulted in a diverse intelligence insight series for different communities—lesbian, gay, bisexual and transgender (LGBT), women, Asian American, African American, and Latino and Hispanic. In this way, Nielsen has developed culturally relevant messages that have moved away from stressing compliance to a message of empowerment through connection with people's culture and heritage.

Nielsen's messaging has three goals: (1) to build awareness and trust through the message and engagement with the community; (2) to develop culturally relevant messages of empowerment, rather than compliance; and (3) to own the message but empower others to share, what Nielsen terms "celebrating the conscious consumer." In all of the communications, Nielsen stresses, in simple and succinct terms, its policy on data privacy.

Lo shared examples of the medium mix used in the messaging. Within the African American advertising campaign, for example, Nielsen found it effective to target high-traffic cinemas where African American consumers tend to watch movies and presented short, digital clips in cinema advertising. In contrast, for the Asian American community, which is oriented to ethnic print media, Nielsen created advertorials—essentially information that the community would want to know about what they are buying and how they are watching television. Nielsen senior leaders wrote some of the most effective advertorials. In other communities, more traditional out-of-home advertising media like billboards and bus wraps were effective in increasing exposure to the Nielsen brand.

Lo explained that external advisory councils were formed to help guide the campaigns within the communities. Nielsen found community-based nonprofit organizations that are focused on civic engagement or voter empowerment are ready to help, and suggested that the Census Bureau use them to get the ACS message out into their community.

In conclusion, Lo made three suggestions for a messaging campaign:

1. Make sure consumers can embrace and own the message. It is important not only to create culturally relevant messages but also to focus on change management from the top of the organization to the field representatives.
2. Stress the importance of response. Change management involves stressing why responses are important and providing sound bites to convince people to embrace the empowerment medium message.

3. Use vignettes. Successful Nielsen vignettes include heritage month videos for each of the diverse communities showcasing the fact that Nielsen knows and is powered by the consumers.

DISCUSSION

In the question period after this panel, one participant asked about targeting brand communication to specific audiences in addition to the general public, such as ACS employees who should be messengers of the value of the ACS. The participant observed that, in his experience, ACS staff view their mission as production rather than sales. The participant asked the panel for comment on how essential it is for the staff in the ACS office to own the brand.

Lo answered from her perspective in change management. The Nielsen advertising and communications strategy addresses ensuring that changes are adopted, that people embrace the message, and that long-term ownership of the message is sustained. Some ways to make sure key stakeholders are part of the solution are to make sure that they are all involved in the process, from the champions who drive the message, to the stakeholders who are leveraging data in the different communities, to the field representatives who are knocking on doors. All stakeholders need to own part of the discussion, she said.

For example, she noted, the field representative guide instructs field representatives to say a simple thank-you and smile. Because people's attention span is so short, Nielsen has incorporated a sound bite for its field reps to not only say thank-you but also indicate the importance of the data to provide insights that will be shared with clients. Bauman added the objective should be to embolden employees to feel like they are brand ambassadors.

The same participant asked about communication to Congress. It was noted that the Census Bureau has a wonderful innovative application programming interface (API) that allows each member of Congress to put Census Bureau information on his or her Website. The API is populated with ACS data by congressional district. The participant urged the Census Bureau to communicate its brand to members of Congress both for legislative purposes and for support of congressional district offices that respond to constituents with questions about the ACS.

A participant observed that Nielsen invests in building a relationship with people who are being asked to respond and that relationship-building has a parallel to the ACS. It implies the construction of a continuous relationship with groups over time. The participant suggested the Census Bureau, which is already going in this direction, should leverage the experience of Nielsen to move forward. Lo responded that the perception Nielsen wants to instill is that respondents have won the lottery, that they have been specially chosen and that they have a responsibility to respond, as consumers who will shape products and services created to serve their needs.

A participant asked if the focus of the messaging should be the ACS or the Census Bureau, observing the agency has huge campaigns associated with the census and the ACS is part of the census. Bauman responded that the Census Bureau has a brand that exists and touches every single household in the country once every 10 years, and the ACS should attach to that brand. She explained this is called brand architecture, where there is a parent (U.S. Census Bureau) and sub-brands. Lo and Terhanian concurred the ACS should capitalize from the efficacies, goodwill, familiarity, and awareness of the Census brand. Terhanian added that building brand awareness is incredibly expensive. The Census brand exists and is very strong, he said, noting he would cobrand the ACS, which is not as strong, with it.

4

Using Administrative Records to Reduce Response Burden

In 2015, the Census Bureau outlined its proposals for actively evaluating alternative data sources, their role, and their quality in *Agility in Action: A Snapshot of Enhancements to the American Community Survey* (U.S. Census Bureau, 2015a). The paper outlined a comprehensive and ambitious program to work to minimize burden for American Community Survey (ACS) respondents and highlighted the need to consider information from alternative data sources, such as administrative records, in place of items on the questionnaire. Further, the paper suggested that the Census Bureau generate information from a merger of responses to any remaining survey questions and the alternative data.

This review had previously been recommended in a National Academies of Sciences, Engineering, and Medicine panel in a report titled *Realizing the Potential of the American Community Survey: Challenges, Tradeoffs, and Opportunities* (National Research Council, 2015). The study panel recommended that the Census Bureau continue research on the possible use of alternative sources and estimation methods to obtain content that is now collected on the ACS. It further recommended that once a comprehensive evaluation of the data needs has been completed for each of the items, the Census Bureau should evaluate whether the survey represents the best source for those data or if data from other sources could be considered as a substitute (p. 10).

In view of this emphasis and at the request of the Census Bureau, the workshop steering committee selected, as one of the four main topics for this workshop, an exploration of how administrative records could replace or improve ACS content. This chapter summarizes the presentations and discussion on this topic based on two sessions of the workshop. Julia Lane (New York University) moderated the first panel, focusing on the Census Bureau and outside expert evaluations of the use of administrative records. Linda Gage moderated the second panel, which included discussion of future directions toward which this initiative might be guided.

USE OF ADMINISTRATIVE RECORDS TO REDUCE BURDEN AND IMPROVE QUALITY

Census Bureau Practices and Research

Amy O'Hara (Census Bureau) provided an overview of what the Census Bureau has been looking at with administrative records involving household surveys. She highlighted three main points of interest in using administrative records with household surveys: reduce burden, make the surveys more efficient, and improve data quality.

Several Census Bureau surveys are now exploring the use of administrative records in an effort to reduce content and, therefore, burden. In addition to the ACS, the research is ongoing for housing items in the American Housing Survey (AHS) and characteristics of individuals and their labor force participation in the National Survey of College Graduates (NSCG). The NSCG is also considering the potential of administrative records to provide information for the periods when the survey is not in the field.

The research focuses on modeling how records could be used to make data collection more efficient. Administrative records are being used to determine the best time of day to reach people and the best mode, O'Hara explained. Also, they are helping identify households likely to have a computer and therefore use an Internet option rather than need a personal interview.

Another use of administrative data is to build sample frames for surveys that have targeted populations. This research centers on the National Survey for Children's Health and the National Teacher and Principal Survey (NTPS). The emphasis for these surveys is to glean information either from federal agencies or, through purchase, from vendors and develop means to incorporate those data into sample frames, mostly the Census Bureau's master address file, which is the Bureau's source for address-based sampling.

O'Hara stated that a final use of administrative records is to help locate people for tracing in the Census Bureau's longitudinal surveys, such as the Survey of Income and Program Participation (SIPP).

Administrative records play a major role in Census Bureau efforts to improve the quality of the information that is collected, O'Hara said. Records from outside sources can help identify underreporting and misreporting, and may help in understanding those errors as well as how to correct for them in modeled estimates.

Administrative records on health insurance have long played this role in the Current Population Survey (CPS) and other surveys conducted with the National Center for Health Statistics. An innovative use of administrative records has been implemented in the SIPP in which information from the Social Security Administration (SSA) that includes earnings as well as SSA payments has been used to impute for missing information in the SIPP. Finally, she said, the Census Bureau is working with the Bureau of Labor Statistics on preliminary research on housing-type variables for the Consumer Expenditure Survey.

Given the important uses of administrative records, O'Hara summarized the authorities and mandates through which the Census Bureau, under Title 13 of the U.S. Code, is authorized to access and use administrative records. The legislation directs the Census Bureau to use administrative data to the maximum extent possible, rather than conduct direct inquiries. Under these authorities, the Census Bureau has been getting data from such federal agencies as the Internal Revenue Service, Social Security Administration, U.S. Department of Housing and Urban Development, Centers for Medicare & Medicaid Services, and U.S. Department of Health and Human Services for decades. The agency has also been going state by state to pursue information on human services programs with person-month detail, primarily from the Supplemental Nutrition Assistance Program (SNAP) and the Special Supplemental Nutrition for Women, Infants, and Children and Temporary Assistance for Needy Families Programs. The aim of these administrative record initiatives is to obtain access to rich sources of information for the household surveys, as well as for the decennial census, on populations that are often hard to count.

Other administrative data are acquired from third-party vendors. Such data include property tax, property value, and deeds information. The vendors add value by aggregating public records on this information and reselling it in a form that can be used by the Census Bureau. Other data sources to consider are the data that the Census Bureau has already collected—demographic characteristics picked up in decennial censuses on race and Hispanic origin and housing structure characteristics from the AHS. For example, a property that is waterfront property in the AHS remains waterfront property in subsequent data collections.

There are a variety of methods through which administrative records find their way into Census Bureau surveys, O'Hara reported. In the SIPP, the information is used in modeling. It is also used through substitution—a method used for the AHS to identify which respondents live in public housing units. Another example of deployment of administrative records is in the frame for the NTPS. In this case, the records did not completely replace information from the survey.

The third method is a hybrid—combining records with the information that has been collected to fill in missing data or to incorporate the records into the estimate in a way that results in an estimate built on administrative data or third-party data as well as on respondent-provided information.

O'Hara reported that the ACS research program has focused on a series of variables (associated with ACS questions) that are seen as candidates for replacement or enhancement with administrative records. The list, published in *Agility in Action* (U.S. Census Bureau, 2015a), was selected on the basis that there was a source of easily accessible administrative data that could possibly have good coverage and good alignment with the concepts on the ACS (see Table 4-1).

O'Hara reported that each of these topics has been evaluated for its contribution to respondent burden (measured in number of seconds required for the answer) and difficulty or sensitivity (as identified in previous research). This project has allowed the Census Bureau to prioritize the most promising variables as the research program moves ahead and to narrow focus on the variables for which there is good concept alignment from an existing data source and that either take a long time to answer or are cognitively difficult or contain information people consider sensitive.

O'Hara highlighted major research studies conducted to date. For one project concerning the "year built" question, the Census Bureau bought a third-party data file that was matched to ACS housing units in the 2012 sample. The match was evaluated on a geographic basis to understand where the year-built data were present from the third-party data. The data did not exist in the third-party files for some of the country. For example, Vermont is completely missing because the data reseller apparently could not obtain the Vermont data. To fill this gap, the Census Bureau would either have to develop an agreement with the counties or develop an open data portal for the state. On the whole, however, there was sufficient coverage in many parts of the country. The quality of the information from the vendor is good because it is a government record, as opposed to the current data, which are obtained by asking ACS respondents about the year their houses were built. The studies comparing the ACS information to that in third-party vendor sources continue.

O'Hara described another recently released study on income that shows a very high correspondence of the ACS data to information observed in the IRS W-2 file. Returns were available for 88 percent of people aged 18 to 64, and the mean wages were within $1,000. The high match rate was even higher for older respondents—returns were available 98 percent of the time for people aged 65-plus. Another study of the availability of housing variables present in third-party files (acreage, property value, and real estate

TABLE 4-1 Priority Topics to Be Studied for Replacement by Data Sources

Topic	American Community Survey Question Number	Estimated Seconds to Complete	Sensitive or Cognitively Difficult?
Phone Service	H8g	1	
Year Built	H2	11	Difficult
Part of Condominium	H16	4	
Tenure	H17	11	
Property Value	H19	11	Difficult
Real Estate Taxes	H20	9	Difficult
Mortgage/Amount	H22a and H22b	11	
Second Mortgage/HELOC & Payment	H23a and H23b	5	
Sale of Agricultural Products	H51	1	
Social Security	P47d	10	Sensitive
Supplemental Security Income	P47e	8	Sensitive
Wages	P47a	41	Sensitive
Self-Employment Income	P47b	8	
Interest/Dividends	P47c	20	Sensitive
Pensions	P47g	8	Sensitive
Residence 1 Year Ago & Address	P15a	18	
Number of Rooms & Bedrooms	H7a and H7b	13	Difficult
Facilities	H8a, H8b, H8c, H8d, H8e, and H8f	6	Sensitive
Fuel Type	H13	14	
Acreage	H4	5	

SOURCE: U.S. Census Bureau (2015a, pp. 8-9).

taxes) also found high match rates between ACS data and data present in the third-party sources.

Despite these promising initial results, O'Hara said challenges exist in using administrative records. The main challenge is data quality. There are questions about how best to assess the quality of the data. The assumption is that a public record is a better source than a survey response, but there are cognitive and definitional differences between the two. It is important to understand the conceptual basis of the sources, she said. There are also coverage issues. As reported above, the administrative data may be missing for some populations and geographic areas.

The matching may also result in errors. The Census Bureau associates the third-party data, the federal data, and the state data to the ACS through probabilistic matching. Although many government files come into the Census Bureau with a Social Security number (SSN), the ACS does not collect SSNs, so the match relies on name, date of birth, address, and with whom the respondent lives. The process used in the matching relies on a key called the Protected Identification Key (PIK), which is a pseudo code that replaces an SSN in order to facilitate deduplication and linkage across the files. For the ACS overall, about 90 to 94 percent of records are matched through the PIK process, but the match rate is lower for some age, race, and Hispanic origin groups. Although the PIK rate does rely on complete accurate identifiers being present in the ACS, the match is difficult when respondents do not have valid SSNs or if they provided the ACS with only their first initial or no date of birth. A match requires that the ACS respondent is present in the Social Security NUMIDENT file.

Even address-matching presents some difficulty, O'Hara said. The ACS frame has addresses (maintained with a master address file unit number) that are based on the physical location of the interview, but administrative data may have post office box or rural route identification. Similarly, property tax records, such as CoreLogic data, refer to the property's basic street address, which does not reflect all units within an apartment building.

Data access is another issue, according to O'Hara. When the Census Bureau acquires information under its authorities, a data use agreement must be executed that states how the information will be used, protected, and destroyed. These agreements need to consider requirements for future, continued access to the information. This is a challenge because the information is quite volatile. For instance, phone numbers have changed because the wireless file has increased with the larger number of cellphone numbers. Vendors have gone out of business or been acquired by other companies. The number of tax filers varies over time with changes in reporting requirements.

O'Hara described two other issues surrounding administrative data to consider. One relates to completeness. For example, some ACS-defined income from the eight-part income question is not reported to the IRS or is reported for periods other than a calendar year. IRS income may not be complete enough to meet the current ACS definition. It might be possible to shift the ACS definition to be more compatible with annual gross income, a household data concept for the primary and secondary filers on a 1040 tax form. There are questions of sufficiency for current users of the current income items. Blending the ACS and administrative data is an option being considered by another Committee on National Statistics panel, and applications of big data are open questions.

Finally, O'Hara posited a series of issues that could affect use of administrative data in place of or supplementing ACS data. It would require a thorough understanding of the characteristics of the new data and assurance of its availability. The Office of Management and Budget would have to approve and a *Federal Register* notice would also be required. The ACS editing and imputation systems would need to be adjusted to accommodate input of other data (now missing data are imputed based on responses that others have provided). In order to assess the impact on the historical data, the Census Bureau would need to simulate the impact on its 1- and 5-year products. Finally, O'Hara said, the Census Bureau would need to make sure of federal agency buy-in because the federal stakeholders for these questions would need to understand the impact of this change of record implementation.

Comments on the Use of Administrative Records to Reduce Burden and Improve Quality

Following O'Hara's presentation, Paul Biemer (RTI International) provided comments from, as he stated, "an outsider's perspective" and focused on reducing burden and improving quality in an optimization context. He identified two possible optimization problems: first, to minimize response burden (the objective function) subject to the constraint that data quality equals or exceeds the quality of the current ACS data, and second, to maximize data quality subject to the constraint that burden is equal to or less than a target level of response burden. In some ways, he observed, these are equivalent. He further observed that to be able to minimize respondent burden, it is necessary to define and measure it and come up with a concept of a reasonable amount of respondent burden. He said valid measures related to data quality are also needed and a concept of a reasonable level of data quality should be developed.

He suggested that one strategy to quell the criticism of ACS burden by Congress and others would be to show that progress is being made on reducing respondent burden. Showing progress requires having a metric to measure burden over time. The measures of data quality could be as simple as tracking the standard errors of estimates (assuming they would be affected by burden mitigation efforts) or an average standard error, or they might include indications of nonsampling errors in order to measure different sources of error or such factors as timeliness.

Biemer referred to the previously discussed article on response burden by Bradburn (1978). From among the list of indicators offered by Bradburn, Biemer suggested choosing a metric or several that permit tracking progress toward a goal. With regard to data quality, he suggested a total survey error approach. Data quality is improved when error is reduced, and the total survey error approach indicates how much data quality is improved. To the list of errors offered by O'Hara, he said he would add specification error, which would include concept alignment and any misalignment of the time intervals.

In addition to specification and coverage error, Biemer would highlight modeling error to identify the impact of record linkage approaches that are modeled or the indirect uses of the administrative data, and within household coverage error, to measure whether information is gathered for all the individuals in the household. Once the sampling and nonsampling errors to be measured are identified, he suggested developing a matrix to portray progress in reducing respondent burden from multiple dimensions.

Use of Administrative Records to Reduce Burden and Improve Quality: A Discussion

Michael Davern (NORC at the University of Chicago) stated that substitution is a viable long-run solution for the ACS to reduce respondent burden, but more immediate solutions to improve quality could be put into place very quickly. In considering immediate solutions, he emphasized the importance of focusing on post-processing actions, since the ACS processing system is extremely complex and interdependent.

One project would be to link data research into data products, providing a restricted-use data product that is regularly updated (annually or biannually) and released with supplemental or additional information that can be linked back to the ACS. The linked data product could be made available to researchers in research data centers and would be very useful in helping researchers improve the quality of the estimates they produce for policy-related purposes. He said it should be well documented, cleaned, and edited, and it should have weights that are created to deal with all the linkage issues, such as cases with missing identifying information.

Davern also supported the creation of blended estimates for public-use files. He suggested that the ultimate goal could be a fully blended or imputed estimate or, at a minimum, simply the model coefficients. He pointed to two substantive examples from a recent administrative/survey data record linkage paper (Davern et al., 2015).

In one study, record-linking research found 22 percent of those in the ACS who are linked to coverage in the Medicaid database do not report having Medicaid in the ACS. Similar findings were the outcome of another linked data research project using the records for SNAP in New York State. Fully 26 percent of cases showing receipt of SNAP in the administrative data did not report it in the ACS. Davern stated his concern about these unreported data because SNAP data are used for important policy purposes. Medicaid and SNAP are important sources of cash benefits used in the supplemental poverty measures. Also, simulation modeling by the

Congressional Budget Office, assistant secretary for policy and evaluation of the Department of Health and Human Services, Centers for Medicare & Medicaid Services, and other federal agencies relies on these data inputs for simulating important policy programs and evaluating whether or not those programs have been successful and met the needs that they were supposed to.

Davern then discussed experimental simulations in which a model was used to create blended estimates. The research has found that use of linked data (administrative and public-use file variables) in a model to impute whether or not people in the CPS had Medicaid or SNAP resulted in an 81 percent reduction in the root mean squared error, mostly due to bias reduction. Although this modeling has not yet been done for the ACS, the results show a significant reduction in potential error with the investment of few resources, suggesting the promise of the approach. Using models also greatly reduces confidentiality concerns, and models can be extrapolated from one geographic area to other areas and from one time period to another. Based on the findings of his research, Davern recommended that ACS data products should incorporate administrative data to reduce burden and improve quality, keeping in mind, however, that incorporating administrative data will tend to affect the time series data because the error structure will change.

General Discussion

A participant agreed that it will be important to have a continuous program of examining the models and looking to update them, because data sources will improve over time, which, in turn, will affect the models that need to be updated. O'Hara added that the Census Bureau administrative records research is changing the way that surveys are being viewed. She further cautioned that there will always be a need for some sort of on-the-ground data collection in order to validate and understand the administrative sources that the Bureau is able to acquire.

Another participant pointed out that the Census Bureau currently uses administrative data for modeling to improve the imputations for program data in SIPP and that administrative data are used in microsimulation modeling with applications designed to improve estimates of cash and noncash benefits. O'Hara added that the SSA is using microsimulation modeling with demographic characteristics from census data in a program that has existed for decades.

Committee on National Statistics Director Constance Citro praised the work that the Census Bureau is doing with the surveys and administrative records, but pointed to findings of a National Research Council report on microsimulation modeling that organizations, such as the Urban Institute and Mathematica, do not generally have access to the full content of administrative records because of confidentiality concerns (National Research Council, 1991). She suggested it would be more useful for the Census Bureau to create the needed modeling infrastructure because only it has access to the full richness of the data.

Another participant raised the issue of obtaining permission from respondents to link survey and administrative records. The participant suggested a need for communication with household reporters when there will be substitution for item nonresponse or wholesale replacement of some answers. Although informed consent does not pertain in a mandatory survey under Title 13, it would be useful to explain the possibility of substitution to respondents, the participant commented.

O'Hara pointed to pages on the Census.gov Website that discuss data linkage activities, and she stated that there are plans to expand those pages. Current outreach materials talk about the Census Bureau combining reported information with other sources. There is an issue concerning whether that information is in sufficiently plain language, she noted. Her organization and others have been working with the Census Bureau's communications directorate to improve outreach and the way to describe the uses of administrative records. New communication initiatives include development of three videos (following the lead of other countries) that explain clearly to the public in cartoons how their data are being used, the benefits to the improved measurements, how the data are obtained, the authority to use them, and the impact on the public.

Andrew Reamer asked about the status of a Committee on National Statistics (CNSTAT) project funded by the Arnold Foundation on improving federal statistics for policy and social science research using multiple data sources and state-of-the-art estimation methods. Brian Harris-Kojetin, CNSTAT study director for the panel, reported on the panel's progress.

Following up on Biemer's presentation of an optimization framework, Greg Terhanian asked if Biemer would develop an algorithm to identify the optimal combination of variables and levels that would produce the optimized survey. Biemer responded that he was not sure if an algorithm could be developed; instead, plans are to compute a metric that measures a definition of respondent burden, and then compute several measures of data quality with the constraint that the data will not be of worse quality than at present.

A participant complimented the optimization framework described by Biemer and wondered about the issue of the relatively rare phenomenon of a person who is extraordinarily dissatisfied with being burdened and who may react in ways that the Census Bureau is concerned about. For the Census Bureau to have interventions that minimize some kind of maximum risk can be inefficient—a form of minimax decision process. It may be better for the phone or personal visit interviewer to terminate the interview rather than antagonize the respondent and end up with little useful information, the participant suggested.

A participant volunteered that a principal driver of using administrative records for the decennial census and other components is to minimize cost. The need to minimize cost should be added as a constraint on the proposed optimization framework. In addition, the participant commented that using administrative records creates a potential nonlinkage bias similar to a nonresponse bias in the survey community. A high linking rate is not necessarily good unless it is accompanied by a measure of nonlinkage bias. For example, for small-area estimation there are benefits of using administrative records, but lower levels of geography could have lower linking rates and lower quality. It would be useful for the Office of Management and Budget to develop standards and guidance on nonlinking bias, the participant suggested. Biemer responded that it would be useful to develop a taxonomy of all of the error sources that are relevant for any particular administrative records and to look at the total error as well as the individual sources of error.

O'Hara countered the Census Bureau's interest in administrative records is based on improving measurement as well as minimizing cost. She agreed that it would be useful to have standards for nonlinking bias. Referring to previous studies based on linking the 2010 census and administrative data, she cautioned it would be difficult to develop standards.

ADMINISTRATIVE RECORDS AND THE ACS: FUTURE DIRECTIONS

This session focused on further uses and definitions of administrative records and future directions for this area of inquiry. The presenters were Julia Lane (New York University) and Frauke Kreuter (University of Maryland).

Rethinking Administrative Data

Lane discussed four topics: (1) the definition of burden in the context of administrative data, (2) the use of administrative records, (3) the sources of administrative records, and (4) possible future directions.

She proposed thinking about the measure of burden as a value proposition. On the cost side, Lane reported the ACS costs taxpayers about $256 million (2017 budget request) and an estimated cost of responding to the survey (respondent time valued at average earnings) is $42 million. Survey error constitutes another cost.

These costs would be compared to the cost of obtaining and using administrative data. The value of the ACS, she said, is the policy value; presentations during this workshop have amply proven its value in the generation of good public and private decision making. However, quoting a 2015 National Bureau of Economic Research working paper (Meyer et al., 2015), Lane said the ACS and other household surveys are "in crisis." The paper documented a massive amount of bias in survey reports relative to programmatic error. As an indication of the declining value of household surveys relative to administrative data sources, Lane referred to a 2012 presentation by Raj Chetty, which reported that the proportion of papers in four leading economic journals that were microdata-based went from about 20 percent to near 80 percent over the past 25 years (Chetty, 2012).

Lane suggested the Census Bureau should adopt a broad view of administrative data. New types of data, including transaction data, are now available that were not available when the ACS was developed 25 years ago. New types of data include cellphone records and data drawn directly from companies' human resource and finance offices. The Census Bureau could adjust administrative data in a much broader context, Lane said, which would include transaction data and camera records and hyperspectral sensors such as are available on Google Street View. Hyperspectral images can be used to determine if people are at home. Microbiome analysis of sewer contents can be used to distinctly identify how many different people are at an address and how often they are there. Likewise, information about commuting and journeying-to-work patterns and mode of transportation can be gleaned from cellphone data.

Unemployment insurance wage records can be used to develop statistics by income earnings and poverty in order to gauge the need for economic assistance, Lane continued. She referred to the Longitudinal Employer-Household Dynamics, a Census Bureau administrative data program with records on all workers in every job in the covered sector and their quarterly earnings. This file is matched with SSA data to provide age, race, sex, and industry with geographic detail to the block level. Occupation can be modeled from job titles in human resource administrative records or from data sources such as LinkedIn, CareerBuilder, and Monster.com.

According to Lane, the problems with nonresponse and missing data can be at least partially overcome with the use of administrative data. Much information can be scraped, analyzed, and predicted from administrative data.

Lane suggested future directions for the Census Bureau's administrative records work. One approach would be to institute pilot projects around high-priority areas such as the transportation workforce. Additionally, she suggested that the Census Bureau build a community that understands what the issues are, that works with the ACS staff to build an administrative records system and that also brings ACS production staff into creating new datasets by conducting training, following the model of the Census Bureau's successful big data training classes. The training could be built around use cases.

Approaches to Implementing Administrative Data

Frauke Kreuter suggested three key points or what she termed rules of thumb for guiding a program designed to increase the role of administrative records as a means to reduce burden: know the inferential goal, dare to combine imperfect data, and empower top-to-bottom teams that work on the issues that have been identified. She elaborated on the three points.

Know the inferential goal Kreuter referred to the work of the CNSTAT panel, chaired by Robert Groves, on integrating multiple data sources and observed that the panel has not yet developed solutions to the challenges of integrating multiple sources. It is a large issue, encompassing many different data products and statistics with multiple uses and different inferential goals. For example, some of these statistical data products, such as the point estimates and other statistics for areas produced from the ACS, are designed for description, but they have acquired other uses such as prediction by third- or fourth-party users of these data. The quality and composition of data best suited for these different uses are very different. For descriptive uses, it is important to have known nonzero selection probabilities, she noted. For the other uses, it is less required to know the selection probabilities. In addressing both sample-based statistics and administrative data, it is essential to know the goal and the unit level. She posed several questions: Is the goal to have data at an individual level? Are microdata records needed for every single household or are block-level or community-level data sufficient? Do individual and household data need to be geocoded? Are the data to be mainly used in generating national estimates or are they to be linked to other data sources?

Dare to combine imperfect data Kreuter ventured that administrative data are imperfect, as has been pointed out, and administrative data will never fully substitute for survey-based data. She further observed that survey, administrative, and found data are all filled with error. Nonetheless, statisticians have experience in combining imperfect measurements in ways that can improve the estimates. She observed that psychologists have developed statistical techniques to combine data and have been able to develop multiple measures for certain constructs.

Empower the ACS team Although experts can help develop approaches, Kreuter stressed that the ACS team ultimately must transform the ACS.

She mentioned, as an example, the Census Bureau's big data initiative. The Census Bureau approach was to develop a class that supported the goal of creating champions at each of the agencies who understand the whole process. The process started with research. The class had a component on data capture followed by data curation, modeling, analysis, output, and ethics. She advocated training programs at universities, government agencies, and in the private sector to create teams for peer-to-peer learning around a data product, and also advocated that ACS should be part of this workshop.

Kreuter urged work on predicting who will respond to the survey and under which approach the person will respond. Statistical models can improve that prediction and machine learning is flexible enough to handle more data. Paradata should be collected and models should be updated constantly, she said.

Machine Learning, Administrative Data, and the ACS

Kreuter also spoke on behalf of Rayid Ghani about a course on big data cotaught by her, Ghani, and Lane. Kreuter highlighted that survey researchers already do machine learning but with different tools, so what is needed is language bridging. Machine learning is an umbrella term for any algorithm or computer program that can learn from experience with respect to a certain task accompanied by performance measures, such as when a credit card stops working due to fraud detection algorithms working in the background that look at patterns, learn from experience, and flag differences. Services like Amazon and Netflix use machine learning to predict what someone will want to watch or want to buy, or not, based on past behavior.

She asserted that the first algorithm taught in computer science and machine-learning classes is logistic regression, a type of machine learning. What is different with these machine-learning algorithms is that they are less robust and less static than other techniques, and they are more flexible and scalable in order to handle much more data.

One of the requirements for using these techniques is to have lots of data available to "train" the model. ACS certainly has a lot of data—including paradata—that can serve as a training set, she pointed out, because the Census Bureau knows whether a household did or did not respond historically, and models can be updated constantly because new data are coming in.

To employ these models, the analyst needs to map what is wanted to predict to the machine-learning problem. There are three different categories of techniques: (1) unsupervised learning, where one does not have a specific area to predict or classify, and the techniques include clustering or principal components analysis; (2) weakly supervised learning, where one has anomaly detection; and (3) supervised learning, where the objectives are classification of a case into one of several discrete types, or regression, where one is predicting a continuous variable.

The steps for implementing these techniques include data preparation, identification of useful features in the data, model building, model validation, and model deployment. A key point on model validation, she noted, is that there are enough data to put 80 percent of the data in the training set and 20 percent in the validation set to determine the extent to which predictions are useful. Unlike many other surveys, the ACS would have the capacity to allow an analyst to do that. There are also a variety of other data sources at the Census Bureau that could be added into these tasks, such as administrative records sources like Longitudinal Employer-Household Dynamics, IRS data, and other federal programmatic data. Clearly, there will also be new data sources, such as GPS data (useful to determine commuting patterns) and video data, she said.

Discussion

A participant suggested that ACS staff rethink the imputation procedures used for the ACS. Currently, if the whole case is missing, the data are generated by a hot deck allocation method. Not only are there no missing data shown, but also there is no flag for indicating whole case or individual variable imputation. She said users need to know if the data have been imputed. In response, another participant reported that the ACS Public Use Microdata Samples (PUMS) does include an allocation flag per variable, which is documented on the Website. However, it is correct that there is no fully allocated flag in the rare instances that a housing unit has fully allocated people. For group quarters, the Census Bureau uses fully allocated people for estimation purposes; these constitute about one-half of the group quarter's records.

The participant also advocated identifying when variables are created either wholly or substantially from administrative data. She advocated for putting a PUMS file in the public domain, perhaps introducing random perturbation for some of the individual data to get around disclosure issues, but noted that the data should be available in outside research data centers (RDCs). The participant asked about legal restrictions concerning the presence of administrative data in public-use datasets.

A participant responded that if the ACS used tax data instead of the income question, there would be restrictions against release of that information. Currently, when the Census Bureau uses IRS data in economic statistics, public-use files are not produced and the data are only available under restricted access in the RDCs. It would be worth exploring creating synthetic data files, grouping variables, or conducting further research to identify what levels of disclosure are permissible, the participant said.

A participant said use of some administrative data (e.g., cellphone usage and state unemployment insurance records) varies by state and by proprietary status. There are coverage issues as well. On the other hand, the ACS is a national dataset, which is consistent across geographic areas and enjoys the same concepts and definitions. Lane responded that all administrative datasets have coverage issues and biases. However, the statistical agencies are able to make sense of those data, assess their validity, and adjust and correct them. Resources should be allocated to the statistical agencies to undertake this work, she said.

A participant asked panel members about research or insights on the different public perceptions of burden between survey modes—self-response or interviewer-provided response. The participant also asked about any research on how people react to the fact that details of their lives are obtained through administrative sources that they are not even aware of and that they know nothing about.

Kreuter agreed that this is an important issue. There are prohibitions against bringing European data to the United States and analyzing them here. This emphasis on privacy is fueled by a lack of trust in government or certain government agencies. In this view, shared data are perceived as burden. However, she said she was not aware of any systematic research that addressed the issues.

Lane added that much data under discussion are already being collected. The challenge is to conduct a test, perhaps in a pilot project, to assess perceptions and the feasibility of using administrative data for these purposes. A participant observed that privacy advocates urge drawing a line between federal administrative data and nonfederal data, as there are different issues in terms of the government accessing federal data or nonfederal data. Privacy advocates say the public has great concerns with the government accessing nonfederal data.

A participant agreed that privacy of data is an issue and pointed out the government collects administrative data for federal, regulatory, and statutory purposes for program administration. Collection raises the question of the proper federal government role in regard to the administrative data. There is concern that the public may view use of these data as a violation of trust. In this regard, the participant asked, should the federal government overlay these data in its databases, or should agencies simply provide ways to link to this other information and let outside researchers and data users do their job?

Lane concluded the session with the observation that these issues have been frequently raised over the past three decades as administrative records have increasingly been employed to improve, supplement, or replace survey data. She stressed the need for pilot tests to assess the potential of administrative records and the issues accompanying their use.

5

Using Improved Sampling and Other Methods to Reduce Response Burden

The lead-off sessions on the second day of the workshop, according to session chair David Hubble (Westat and member of the steering committee), were designed to lay the foundation for matrix sampling, discuss modeling and imputation associated with matrix sampling, and evaluate the implications of this technique for reducing burden.

CENSUS BUREAU RESEARCH ON MATRIX SAMPLING

Mark Asiala (Census Bureau) focused on a report written by a group of Census Bureau staff who looked at the feasibility of using matrix sampling or other techniques to reduce respondent burden. His discussion covered highlights of the report, options considered, examples of how one might implement these options, statistical challenges identified, and recommendations that came out of that report (U.S. Census Bureau, 2015b). The motivation for the report was to consider means to reduce respondent burden. The review began with consideration of the findings of the Census Bureau's 2014 content review (U.S. Census Bureau, 2014), which had several phases. The first phase was to justify each question on the American Community Survey (ACS) by looking at the frequency that the estimates for that particular topic were needed, the level of geography that was needed, and the legal justification for the use of that question by the various federal agencies. In essence, he said, the first phase was intended to identify any candidate questions for removal. The second phase, which led to the feasibility study, took the rich database developed for the first phase, which had information on every topic and the frequency it was needed. This phase assessed whether the questions could be asked of fewer respondents on only a subsample of the forms or if there were other means of reducing burden short of completely removing a question from the form.

The internal Census Bureau team that did the investigation was composed of people from many different areas within the agency, including those with expertise in operational aspects and statistical aspects, as well as subject matter experts who brought their understanding of the data and how they would be tabulated and used. The team developed four different options for the ACS (see Box 5-1).

Asiala explained that the first option—periodic inclusion of questions that were not required every year—would mean that questions could be asked some years and not others and thereby burden could be reduced.

Options 2 and 3 were variants of a matrix-sampling approach. Option 2 would target matrix sampling for a small set of questions—those not needed for production of small-area estimates that could be asked less frequently. Option 3 was a more aggressive approach to matrix sampling in which a broad set of questions would appear on only some questionnaires and not on others, yielding more dramatic improvements in reducing respondent burden.

Option 4 was a hybrid approach that would use administrative records to fully substitute for survey data in those areas where the records were

---

BOX 5-1
Options for Further Study to Reduce Respondent Burden

- **Option 1. Periodic Inclusion of Questions:** including questions on the ACS questionnaire only as frequently as the mandatory and required Federal data uses dictate
- **Option 2. Subsampling:** customizing the questionnaire such that the sample for individual questions is designed to meet the geographic need specified by the Federal uses of the resulting data
- **Option 3. Matrix Sampling:** dividing the ACS questionnaire into possibly overlapping subsets of questions, and then administering these subsets to different subsamples of the initial sample
- **Option 4. Administrative Records Hybrid:** using alternative data sources as a direct substitution for survey data collection, potentially in a hybrid approach by including the question on the survey in certain geographic areas to address coverage gaps in the alternative data, or to assist in periodically refining statistical models that use the administrative records to meet data needs

SOURCE: U.S. Census Bureau (2015b, p. 1).

strong. When records coverage was low or the quality did not meet an established threshold, the question would remain on the form.

Two criteria were used to identify the topics considered for options 1 and 2. The first was that there were no required or mandatory uses in the law or case history that required use at the Census tract level, although it could be needed at the county, state, or higher level. The second criterion was that all mandatory or required uses were needed at a frequency that was less often than every year. The team identified topics as potential candidates for the matrix sampling. These topics are subject to further verification with the individual federal agencies, but they provided good examples to consider in a feasibility study.

Asiala laid out each of the options in greater detail, giving examples of how the approaches would be operationalized. The options appear in the following illustrations, shown as Figures 5-1 through 5-4.

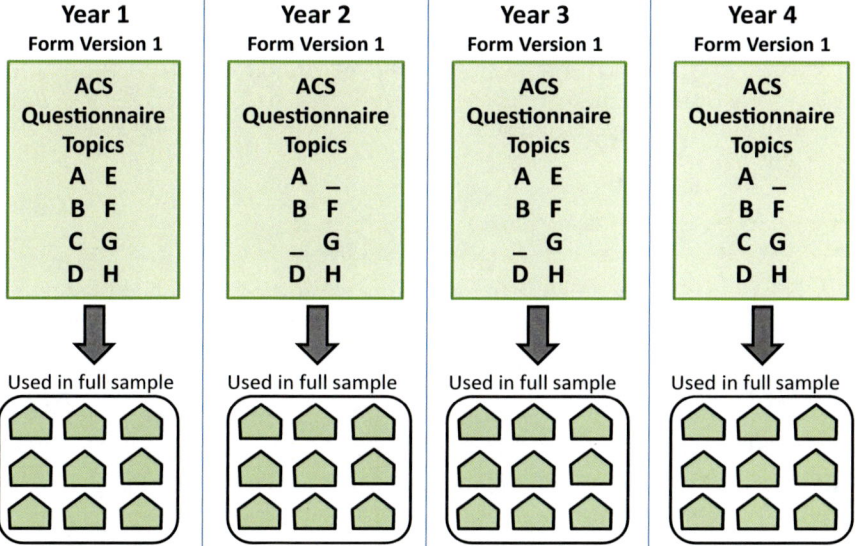

FIGURE 5-1 Illustration of option 1: Periodic inclusion.
SOURCE: Mark Asiala presentation at the Workshop on Respondent Burden in the American Community Survey, March 9, 2016. Available: http://sites.nationalacademies.org/cs/groups/dbassesite/documents/webpage/dbasse_173161.pdf [September 2016].

In this example, Topic B is needed only at the state level, while Topic D is needed at the county level.

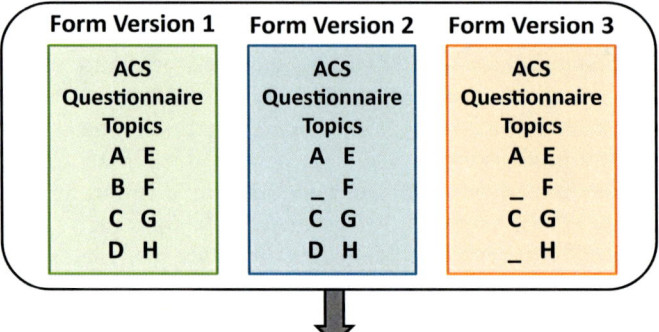

Only the portion of the full sample that is needed to produce estimates at the necessary geographic level receives the corresponding form version. In this example, only a small subset of housing units get Topic B while most, but not all, get Topic D.

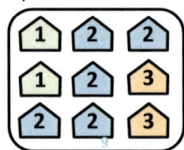

FIGURE 5-2 Illustration of option 2: Targeted matrix sampling.
SOURCE: Mark Asiala presentation at the Workshop on Respondent Burden in the American Community Survey, March 9, 2016. Available: http://sites.nationalacademies.org/cs/groups/dbassesite/documents/webpage/dbasse_173161.pdf [September 2016].

For option 1, in which some questions would be periodically included, a topic could be included in year 1 and year 4 but not in the years in between and another in years 1 and 3, but not 2 and 4. This option would mean that there would be only one form each year. The Census Bureau has experience with switching from one form to a different form by year.

Asiala contrasted this approach with option 2's targeted approach utilizing multiple questionnaires in a given year. For example, if a topic is needed only for estimates at the state level, it could appear in form version 1 but not in versions 2 and 3. If a topic is needed for estimates at the county level, it could appear on form version 1, which includes all variables, and on version 2, but not version 3. In this option, the frequency with which questions appear on the different forms would be determined by the required reliability needed for the particular topics.

Option 3 takes option 2 further, Asiala stated. Rather than consider-

In this example, topics are assigned to form versions in a partially overlapping manner.

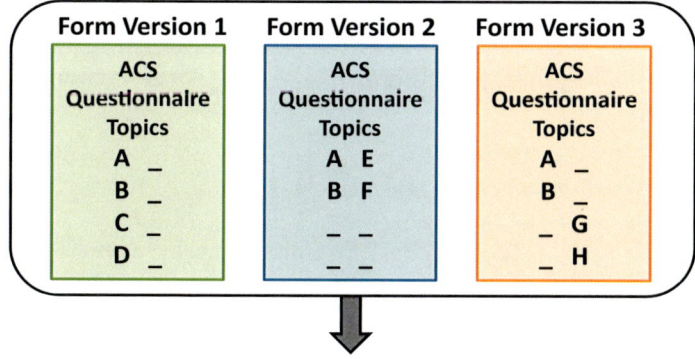

Use either statistical tools or an increase in total sample to help mitigate the impact of the missing data.

FIGURE 5-3 Illustration of option 3: Comprehensive matrix sampling.
SOURCE: Mark Asiala presentation at the Workshop on Respondent Burden in the American Community Survey, March 9, 2016. Available: http://sites.nationalacademies.org/cs/groups/dbassesite/documents/webpage/dbasse_173161.pdf [September 2016].

ing a narrow set of topics, all characteristics or all questions would be available for matrix sampling. Option 3 could mean, for example, that only some topics appear on every form, while other topics appear, on only one of three forms. A drawback of this option is some loss in reliability if adjustments were not made. One option could be to use statistical tools to mitigate some of that loss in reliability. Another option would be to expand the initial sample in order to increase the amount of information gathered from respondents about particular topics.

Option 4 is similar in that an alternative data source would be used except in areas with insufficient coverage. In those areas, the traditional form would still be used. If, for example, an allocation shows administrative records could be used for five-ninths of the country and four-ninths would still need the form, the reduction in burden would be nearly half for those questions.

In this example, Topic G has an alternative data source with good quality and coverage for most geographic areas that can be used directly in place of collecting the data on the questionnaire in those areas.

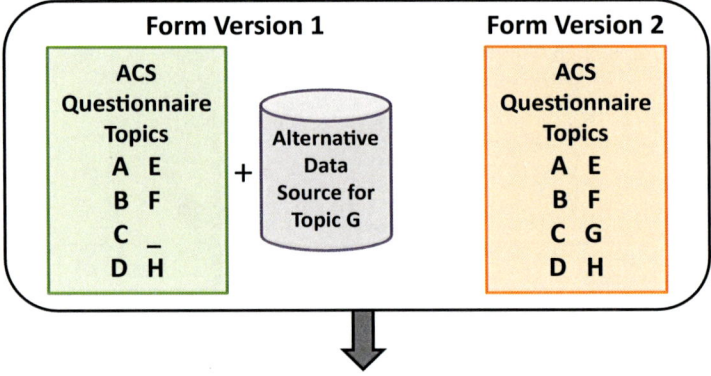

Areas with good coverage for the alternative data source receive form Version 1, while areas without good coverage for the alternative data source receive form Version 2.

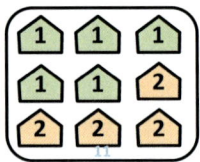

FIGURE 5-4 Illustration of option 4: Administrative records hybrid.
SOURCE: Mark Asiala presentation at the Workshop on Respondent Burden in the American Community Survey, March 9, 2016. Available: http://sites.nationalacademies.org/cs/groups/dbassesite/documents/webpage/dbasse_173161.pdf [September 2016].

Asiala reported that the team adopted a number of criteria to identify the options that were viable candidates for more research. The criteria include operational and processing complexity, statistical complexity, impacts on reliability and accuracy of the data, data availability, the richness of the resulting data products, and a nominal sense of the expense of an option. The ratings—high, medium, and low—were based upon the team's professional judgment, taking into consideration operational, statistical, and subject matter considerations.

He reported that the challenges of option 1 were mainly operational, but changing from one form to another already happens in the ACS on a regular basis and this option would perhaps make it more regular. The team concluded that this option applied only to a small set of topics and, for that set of topics, the changes could be instituted comparatively easily in contrast to the other options.

The team also concluded that for the administrative records option (option 4), there were a lot of topics for which administrative records could potentially be used so the potential for burden reduction is significant. However, in order to select this option, much groundwork to investigate the quality of those administrative record sources and the appropriateness of the use of those administrative record sources would be needed. This option is something that the Census Bureau should continue to explore, Asiala said.

The key challenge identified for matrix sampling is that there will be holes in the data in cases where a question is not asked and there is no response. It would be possible to do cross-tabulations in the traditional sense of using only data from respondents who were asked all questions involved in the cross-tab, but using data only from this limited population would result in considerable loss in precision in the estimates, he said. The team supported creating a complete microdata file unless some other statistical means for accomplishing cross-tabulations would be feasible. If the Census Bureau does not create a complete microdata file, this would introduce concerns about the user friendliness of the public-use file. On the other hand, constructing a complete file introduces issues with the techniques for imputing the missing data. The issues include how to deal with the potential loss of precision of the estimates, the tools that could be used to mitigate some of that loss, and, since the ACS publishes a margin of error for every estimate, how to properly reflect that variance due to imputation in the variance estimates published.

Asiala reported the results of a literature review to find examples of other large demographic surveys that use matrix sampling and to identify statistical methods that such surveys use in their implementation. The team found a number of simulation studies and proposals for matrix sampling for particular surveys. These studies generally concluded that matrix sampling could be done with relatively low impact on bias and that it was possible to mitigate a certain amount of the reliability impact. However, the team did not find any good examples of surveys using matrix sampling in ways applicable to the ACS.

The literature search also considered estimation approaches that relied on imputation techniques to fill in holes, including both multiple imputation (Raghunathan and Grizzle, 1995; Thomas et al., 2006) and hot deck techniques (Gonzalez and Eltinge, 2007). In addition, literature discussing best linear unbiased estimators (Chipperfield et al., 2013) and generalized regression (Merkouris, 2015) was reviewed. One takeaway was that the papers stressed the importance of optimizing how topics are grouped by the different forms that get used in the matrix sampling. Asiala concluded that grouping has a substantial impact on the efficiency or the productivity of these various methods.

The team's empirical explorations were geared primarily toward option 2 (targeted matrix sampling) with the aim to identify the amount that a subsample reduces burden for topics that do not need tract-level detail. To test the effect of subsampling, the team started with the nine topics identified, five of which were required only at the state level and four at the county level. The team then estimated the potential reduction in sample for the target geography and a target coefficient of variation (CV), concluding that the extent to which a topic could be subsampled depends on its prevalence and the geography.

Using a target CV of 10 percent for a state-level estimate (generally considered to be reasonable), the team could achieve a subsampling rate on average across the country of only about 25 percent for low-prevalence items. However, for high-prevalence items, the team achieved substantial reductions—the equivalent of a 1 in 20 sample or a 95 percent reduction for those higher-prevalence items.

At the county level, the team looked at target CVs of 10 percent, 20 percent, and 30 percent. The research confirmed that some counties currently do not have a CV of 20 percent for the tested characteristic so subsampling is not possible for that group. However, larger counties can be subsampled, and a reduction of about 60 percent could be achieved.

From this Census Bureau research project came a number of specific recommendations, which Asiala summarized as follows: (1) implement Periodic Inclusion (option 1) wherever possible given relatively low operational and statistical complexity; (2) assess administrative record sources (option 4) that could be used to partially remove questions from the form, recognizing that this option could translate to significant reductions in burden with relatively few undesirable impacts; and (3) for matrix sampling (options 2 and 3), recognize that there are potentially large impacts on costs and also negative impacts on accuracy and richness of survey estimates, and that the ACS would seek input to help develop research into efficient and effective designs for use of matrix sampling.

UTILIZING MATRIX SAMPLING TO REDUCE RESPONDENT BURDEN

Two discussants, Jeff Gonzalez (Bureau of Labor Statistics) and Steve Heeringa (University of Michigan), followed Asiala's presentation on Census Bureau research on matrix sampling.

Gonzalez said he would take a broad perspective on utilizing matrix sampling to reduce respondent burden, provide a definition of matrix sampling, discuss the design of matrix samples, identify implications of simple matrix-sampling designs, suggest design features that should be considered in conjunction with matrix sampling to achieve further reductions in burden, and highlight statistical and operational considerations when implementing a matrix-sampling design.

He said the motivation for matrix sampling comes from a body of literature on survey methodology that suggests that high respondent burden, low survey response rates, and questionable data quality may each be associated with lengthy surveys. One possible solution to address issues related to burden while improving data quality and nonresponse properties of the survey is to administer a reduced-length questionnaire. However, it is difficult to eliminate questions from the original questionnaire because stakeholder needs vary and rarely are there expendable questions on surveys. This has led the ACS to consider the possibility of dividing the lengthy ACS questionnaire into subsets of questions and then administering each subset to subsamples of the full sample. This is referred to as matrix sampling, he explained, but it is also referred to as a split questionnaire design—these designs ensure that every question is administered to at least some portion of the sample.

Gonzalez offered several illustrative examples of matrix-sampling design, each with special features to consider when meeting various survey objectives. In the six designs in Figure 5-5, the shaded square represents collected data; the open squares represent data missing by design. The rows denoted by S_n represent subsamples of the full sample, while the columns denoted by Y_k represent a specialized subset of questions.

The subset (a) at the top row, left position, represents a matrix-sampling design in which each subset of the full sample receives one specialized subset. Gonzalez stated that this design is adequate for estimating univariate statistics from the portion of the sample that receives that specialized subset, but is unlikely to satisfy the complex data requirements for the ACS. The subset (d) has the specialized subset feature with two additional features—a full questionnaire subsample, which is a portion of the sample that receives the full body of questions, and a common core set of questions that every sample member receives. This design, according to Gonzalez, could satisfy a broad range of data products for the ACS. He discussed other options, including subset (f), which depicts a completely unstructured matrix sample design in which the subsets vary dramatically by the sample members.

While the literature suggests that matrix-sampling designs can improve the overall quality of a survey and perhaps reduce burden, he observed, there are potential disadvantages of implementing these types of designs. First, there is a loss of information obtained in the survey by creating incomplete or missing data by implementing these types of designs. There is also the potential to introduce additional sources of variation such as measurement errors due to context effects. In addition, there is a reduction

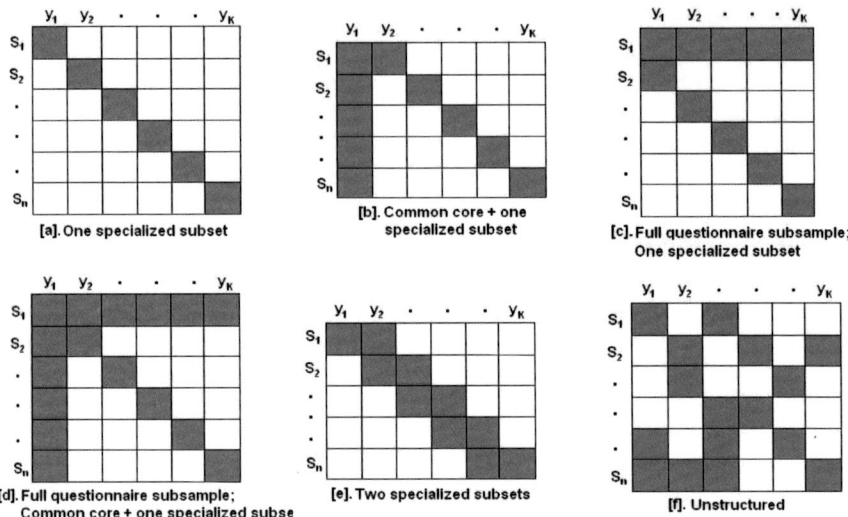

FIGURE 5-5 Illustration of matrix-sampling designs.
SOURCE: Jeffrey Gonzalez presentation at the Workshop on Respondent Burden in the American Community Survey, March 9, 2016. Available: http://sites.nationalacademies.org/cs/groups/dbassesite/documents/webpage/dbasse_173170.pdf [September 2016].

in the precision of estimates from those questions that are matrix sampled. Finally, survey operations are more complicated and case management systems must be modified in order to keep track of the various forms.

Balanced against these disadvantages, Gonzalez referred to work on matrix sampling or split questionnaire designs by Chipperfield and Steel (2012) that identifies three key benefits: (1) sample size requirements to meet survey objectives often differ by survey items, so it makes intuitive sense to allow that difference in survey data collection procedures; (2) leveraging information can enhance design and analysis; and (3) reducing the length of questionnaire through matrix sampling has the potential to reduce burden.

Gonzalez then turned to a further definition of burden in order to assist in exploring how matrix-sampling designs can be utilized to reduce respondent burden. As discussed earlier in the workshop, burden is a perception of the respondent. It is multidimensional in that it concerns not only length of the survey, but also other factors, such as the effort required to respond to the questions, the difficulty of the questions, the sensitivity of the questions, and the frequency of being contacted.

With that background, Gonzalez discussed how matrix-sampling designs can be used to reduce respondent burden. The key question is, given this multidimensionality of respondent burden, how can the Census Bureau or other national statistical organizations design matrix-sampling forms to reduce respondent burden? How can the impact in terms of burden reduction be measured? He observed that simple implementations of how matrix sampling reduces burden can be measured via objective criteria, such as the number of questions or the length of the interview, but, he asked, how are the other dimensions measured? He posited that three substantive issues arise when addressing other dimensions of burden: (1) how to allocate survey items to forms and then forms to subsamples of the full sample so as to improve on those other dimensions of respondent burden; (2) what additional design features to combine with matrix sampling to achieve further reductions in burden; and (3) the extent, if any, that burden reduction affects quality and precision.

With respect to the first issue, the typical design of a matrix sample is such that survey items are allocated to forms and then randomly distributed on those forms to subsamples of the full sample. Previous research has tested various allocations of items to forms, including random allocation. The research has assigned questions to matrix-sampling forms without regard to any characteristics of the questions, permitting item stratification methods or correlation-based methods in which highly correlated items are placed on different forms to be employed. Gonzalez stated the literature suggests that the ability to address dimensions of respondent burden other than length can be affected by this allocation process. For example, employing a technique like item stratification can ensure that forms are balanced with respect to stratification classes where the strata are formed by sensitivity, effort, or difficulty.

Furthermore, Gonzalez said, the greatest impact on burden reduction may not be achieved with random distribution of forms to subsample members because auxiliary information about the sample unit is often ignored. Since the ACS collects information from heterogeneous target populations, the survey methods literature concludes that collecting information on heterogeneous target populations from standardized instruments may be suboptimal. Incorporating auxiliary information about sample units in the ACS can have positive effects on data quality and burden reduction.

Gonzalez said a second implication of the typical design is that matrix samples are ineffective for rare-event items and small geographic or other domains, often because to obtain information for rare items or small geographic domains, questions are added to the set of questions that every sample member receives. If there are no questions eliminated from the original questionnaire, there is no reduction in burden.

In order to achieve better outcomes with respect to respondent burden,

Gonzalez advocates the use of combining or other design features in conjunction with matrix sampling. The other design features include responsive or adaptive designs. They are analogous to multiphase designs and require midcourse design decisions and unit-level survey changes based on the accumulating process in survey data. The basic motivation for considering these designs is that the decisions are intended to improve the cost and error properties of the resulting statistics. In the context of the ACS, the multiple phases are the follow-up procedures of initial or soft refusals.

He stressed that adaptive design decisions require the availability of useful auxiliary information, an understanding of how that auxiliary information can be modeled to determine the best type of matrix-sampling design, and knowledge about how to combine the multiple phases of data collection to produce desired estimates. These topics are being actively researched. He added that an adaptive or a responsive design in conjunction with a matrix-sampling design has the potential to tailor the interviewing experience to the sample unit in a manner that may increase motivation with the survey request. The theory is that this method would elicit more thoughtful and thorough responses from the respondent. Furthermore, by incorporating auxiliary information, it may be possible to mitigate frustration or concerns of inconvenience when respondents feel that they have been contacted too many times or have spent too much time on the response task.

The second design option under consideration involves the integration of data sources, which would comingle ACS or other survey data with information provided in other data sources such as administrative records, other surveys, or organic data sources to either replace, edit, impute, or use in estimation in some way. The motivation for considering these methods is similar to that of matrix sampling, that is, to reduce respondent burden, improve some dimension of data quality, and yield some cost savings. There are, however, concerns with integrating data sources dealing with access to or capturing those alternative data sources, quality, and the ability to link the alternative data sources to the ACS data. Although there may be concerns with privacy in the use of administrative records, Gonzalez observed, respondents already think government agencies share information. Also, when a respondent provides consent to link to the survey data, the respondent benefits from the reduced perceived time spent completing the survey request.

Gonzalez cautioned about additional statistical and operational concerns to consider when deciding to implement matrix-sampling designs. Matrix design introduces data collection issues because, as the number of matrix-sampling forms increases, data collection management gets more complicated. Modification of processing systems is required to keep track of completion rates by forms. Also, each form may have a differential error property, and there might be differential error properties of the same forms across different modes of data collection that are used in the follow-up procedures as well.

A second set of issues deals with response tasks or cognitive issues, he said. There may be context effects in that responses to questions can be affected by prior items administered in the questionnaire, which may provide cognitive cues to the respondent.

A final set of issues relates to data production and analysis, he said. Implementation of a matrix-sampling design will require modifications to the data processing systems no matter whether imputation, weighting, or modeling is used, and it will require a large investment to modify the current systems. The key point is that as the number of forms increases, processing complexity increases and missing data patterns arise.

In conclusion, Gonzalez stressed tradeoffs among burden reduction, total survey quality, and costs. He suggested several high-priority questions for an expert panel to consider when deciding on implementing a matrix-sampling design for the ACS. The first is the meaning of burden reduction—remembering that burden is not simply the length of the questionnaire but includes other dimensions such as sensitivity of questions and difficulty of the response task. The second question relates to the additional design features that should be considered in conjunction with matrix sampling to achieve greater reductions in burden and improve overall survey quality and maintain or reduce survey costs. These include responsive designs and integrating data sources, but there are other features that could be considered with matrix sampling. The third question relates to how the existing ACS data, either survey data or paradata, can be used to inform the matrix-design process. Decisions must be made about allocating survey items to forms and distributing those forms to subsamples of the full sample, modifying the design to account for soft or initial refusals, modifying the matrix-sampling design to account for those soft refusals, and developing criteria to evaluate the new design features.

PLANNED "MISSINGNESS" DESIGNS AND THE ACS

Heeringa focused on planned "missingness" designs, a term he said he prefers to define a broad class of designs that includes matrix samples as a special case. He stated he would cover burden and information needs in the ACS; research-based results, empirical findings, and common-sense observations on planned missingness data designs; four options for incorporating planned missingness or matrix sampling in the ACS; and methodological and empirical issues that are involved.

He explained that the burden of the ACS can be classified as individual respondent burden, aggregate sample burden, system (data producer) burden, and data user burden. Each of these types of burden is driven by information needs, which are, in turn, defined by time, content, and spatial and other domains of analysis, such as subpopulations. Each of these information drivers can be collapsed to reduce burden. For example, the ACS produces data on tracts and on block groups only once every 5 years and develops annual estimates only for areas with a population above 65,000. Linking the drivers to matrix-design issues, he observed that the ACS is already a matrix design over space and time.

He pointed out that the literature discusses three major approaches to design—multiphase sampling, split questionnaire design, and a hybrid of the two. Multiphase sampling was proposed by Navarro and Griffin (1993), who examined potential uses of matrix sampling and related techniques in the 2000 census, and has been enhanced by combination with small-area estimations (Gonzalez and Eltinge, 2010). The 1970 census had this type of a design with a 15 percent sample (Form 2 items) and a nested 5 percent sample (Form 1) of households completing the larger set of questions in addition to the Census short form, and it has proven quite effective over the years. Other examples are the National Health and Nutrition Survey and the National Comorbidity Survey Replication.

The split questionnaire design (SQD) is a pure matrix-sampling design and is usually exemplified by a set of core questions and then modular components. In contrast to the multiphase design, which creates a monotonic missing data problem, the SQD creates a generalized missing data problem. In Heeringa's assessment, the SQD works in surveys and measurement settings where there are many measures that can be modularized into correlated sets of items or blocks of items; for this reason, it is often used in educational testing or educational measurement.

Heeringa concluded that the SQD works when the full survey process is designed for a split questionnaire design or redesigned from the ground up. He referred to a research project for the National Children's Study to determine whether matrix sampling might be a solution to the study's serious problem of cost and complexity. The conclusion of that study was that a split questionnaire design or a split measurement design was not feasible because the survey already had built a large infrastructure, large expectations, and a large interest in diverse sets of items.

He further emphasized that split questionnaire designs work best when there are descriptive and predominately univariate estimation problems. He referred to a research project to develop norms for tests in the president's physical fitness testing program (President's Council on Physical Fitness and Sports, 1985). The challenge was that kinesiologists were changing the set of tests that children were taking from one period to the next, so the research had to set norms for new activities as well as calibrate them with the old for a long list of activities. The data were collected in a national probability sample of classes and schools. The solution was to break the tests into three modules and assign each child or each classroom two of these modules. The method yielded distributions for each test on two-thirds of the sample, and correlations between modules based on two-thirds of the sample if the tests were in the same module and in one-third of the sample if in different modules. According to Heeringa, the method was successful because it started from the ground up and involved a simple set of measures that were correlated and could be modularized.

He offered three suggestions for estimation and inference for planned missingness designs:

1. Full measurement on a subsample of the entire sample strengthens the precision of estimates and the coverage properties of the intervals because it yields a training set of data to inform the full multivariate relationships among the variables. While it does not reduce the burden for the people who have to complete the full set, it does reduce aggregate burden and improves statistical precision.
2. Core content should always be included for all individuals as opposed to simply having random modules assigned to individuals.
3. Regardless of the strategies—multiple imputation, EM, or FIMI estimation—the fraction of missing information will be smallest when the modularized content has high correlations between the modules and heterogeneity within the modules.

Heeringa offered a series of ways that a planned missingness approach could help the ACS. He supported the periodic inclusion of questions (option 1 as described by Asiala) because the ACS program has experience with questionnaire changes at the start of the year. Under this option, he proposed annual estimates based on 1 year of data for the nation, states, and places of 65,000 population and over, and that for those periodic annual estimates, standard weighting and estimation would apply. He proposed an alternative approach to the current practice of collapsing over time (i.e., 5 years) to support spatial estimates by introducing a 5-year rotation of annual topical modules, noting that collapsing over time (i.e., 5 years) yields a five-module split questionnaire design. He did not support split questionnaire designs on an annual basis because the analysis would be very complex. His recommendation was to modularize by year and treat the 5-year interval as a split questionnaire design.

Heeringa also addressed a series of methodological and empirical issues in applying planned missingness in the ACS. It is important, he said, to be mindful that too much modularization may lead to unanticipated context effects in the questions. The planned missingness approach would require the Census Bureau to optimize and streamline processing all through the data handling, data acquisition, form assignment, imputation, and weighting steps. The estimation step would be enhanced by utilizing composite estimation and other small-area estimation methods that borrow strength by blending small-area model-based estimates with direct survey estimates. Other survey aspects, such as cross-tabulations, would require rethinking, and custom tabulation systems for geographic units would need to be developed. Finally, ACS products such as Public Use Microdata Samples (PUMS) and analytic datasets would need attention, and the Census Bureau may wish to consider implementing a user-managed approach to the planned missing data.

GENERAL DISCUSSION

Workshop steering committee cochair Joseph Salvo observed that the 1970 census started from a strong base—the long form had a substantial sample with items asked of 5, 15, and 20 percent samples. The ACS starts from a different point. Its small-area data samples at the tract and blocked group levels have high levels of sampling variability. In his view, introducing a split questionnaire could affect the quality of the small-area estimates. Heeringa replied that small-area estimates are important. He summarized recent experience with using models to improve estimates of the foreign-born population by tract level. He said the results were encouraging: the estimates of the foreign-born after 1990 for every block group when summed to the tract level matched very well (within the limits of sampling error already imposed on these block group level estimates). This approach borrowed strength from relationships that had been measured more accurately at a higher level and assumed that those relationships held with appropriate covariate control.

A participant with experience in matrix sampling in an education application observed that the primary strength of this methodology is that measuring education proficiency in math, science, and other subjects yields a list of items, not a single item like the ACS. The ACS is interested in individual variables, not necessarily a summary level. The education context may not be necessarily compatible with the ACS context. Asiala responded that the real power of the ACS comes from cross-tabulations and looking at the relationships among variables.

MODELING AND IMPUTATION

This session, chaired by John Eltinge (Bureau of Labor Statistics) featured presentations and discussion on modeling and imputation and on maximum likelihood approaches to compensating for missing data.

Modeling and Imputation Discussion

Michael Brick (Westat) continued the discussion on matrix sampling, noting the process has been used in many educational settings. An early childhood longitudinal study uses matrix sampling, in that it samples from items associated with cognitive abilities and math abilities to produce a score for the child. Those scores are correlated with other individual characteristics in order to assess progress over time at the individual level. This is a very different type of setting from the ACS, he noted.

He also observed that the length of the questionnaire has been used as a measure of burden for a long time, but it is not the right measure. The right measure is what respondents feel about what they have just been put through, but that is a hard thing to measure. Thus, measuring length becomes the usual metric.

He made several suggestions regarding practical implications of improving sampling and estimation processes with a missingness design for the ACS:

- **The design processes should be simple.** Any changes will fail if users are not satisfied, he asserted. Most users want to get a table about their area, and they want to run it and be able to present it clearly. If it is not simple, the ACS will not be used as often as it currently is being used. For example, because mass imputation is associated with any type of missingness design—whether multiple imputation, fractional imputation, or hot deck imputation—there is a risk of misleading readers about how much information is associated with those data. The casual user who produces tables and cross-tabulations may have difficulty, he said. The way around this is to plan the analysis and impute for the planned analysis, which takes a sophisticated user. As it is, sophisticated users need more ability to get to the data and use them in a reasonable way.
- **The analysis should be consistent with changes in the data collection procedure.** Brick advocated for coordination between design and analysis, although he acknowledged the difficulty because the ACS has been conducted in much the same way for so long. Some changes will call for starting again with new systems that are appropriate to the new design, rather than trying to tweak the old systems. The data files must be restructured as well, he said.
- **Only data collected on the ACS form should be labeled ACS data.** Administrative data should clearly be labeled as such, and he stressed that linked data are not ACS data. This is needed because the quality of administrative data will change over time and space as organizations that collect and process the data change. An example of how to treat non-ACS data is the phone-service item. Many

commercial organizations produce data on phone service, and Census Bureau analysis makes clear that the commercial data are not the same as the ACS data. Imputing or substituting the commercial data for the ACS could be misleading and would not, in the end, reduce burden. However, administrative data have a huge role in imputation of ACS data, and the Census Bureau needs to allow users to link the administrative data in an easy way.

Matrix Sampling, Maximum Likelihood Approaches, and Multiple Imputation

Paul Biemer (RTI International) began his discussion with reference to the National Longitudinal Survey of Adolescent to Adult Health (Add Health), for which he serves as one of the statistical directors. When the survey converted from personal interviews to Web and mail collection, there was consideration of breaking up the questionnaire like a matrix sample. The program decided against developing a matrix sample due to the fact that it is a longitudinal survey and the matrix sample would create missing variables on the last wave and increased complexity. The Add Health solution was to break the questionnaire into two parts and treat them sequentially. This fall the program will experiment with that approach, testing the full questionnaire versus two modules given to the respondents sequentially, with incentives to complete both.

Instead of matrix sampling, Biemer suggested other options. For example, the Census Bureau might consider multiple imputation. However, since analytic results from multiple imputation may be affected by the imputation model, he posited using a full information maximum likelihood (FIML) approach. He said that this approach is an alternative to imputation because it compensates for missing data; however, it does not replace missing values with imputed data. The records containing missing values are retained for the analysis by putting the missing data mechanism in as part of the likelihood for the data analysis model. He stated that Mplus users (and users of a package called Latent Gold) are used to dealing with these multiple equation models, and the missing data mechanism model would just be another variation.

Biemer listed four advantages with using FIML in the modeling: (1) the method does not fill in missing items but adjusts the estimates; (2) it is more efficient than multiple imputation to which it is theoretically equivalent; (3) because it is a maximum likelihood approach, the standard errors are correct; and (4) it solves the problem of how to impute for a categorical variable. However, the method also has a disadvantage in that it assumes a probability distribution (because it is a likelihood approach). While that characteristic is not a problem when dealing with categorical data because multinomial distributions are usually assumed and those assumptions are not difficult to satisfy, for continuous outcomes it is necessary to assume normality. Another disadvantage, he said, is that it becomes a more complex type of modeling if more complex missing data mechanisms are used. The missingness might depend upon some variables in the core, or it could be some variables in the modules themselves.

He then laid out a simulation of maximum likelihood methodology in practice (see Figure 5-6). The simulation contained two questionnaires, both with a set of core variables and a module and two samples.

For each of the modules, only half the sample is available. For the core, a full sample is available. By modeling this with the FIML approach, it is possible to take advantage of the correlations between the core in module A and the core in module B to create point estimates that have better precision. Likewise, correlations between A and B can boost the precision of the estimates from module A.

Biemer continued his simulation to develop two subtables, crossing C (an item from the core that is given to 100 percent of the sample) with A (an item from module A) and B (an item from module B), as depicted in Figure 5-7.

In the absence of information on A × B, it is not possible to produce that subtable, but the subtable can be obtained by cross-classification information through FIML. The cell numbers would be estimates based upon a model, and how good they are depends upon how well the model fits. The notation in this table is unweighted, but weights can be inserted using packages such as Mplus, Latent Gold, or, as Biemer used, LEM.

Biemer described how this works in notation (see Figure 5-8). Based on an approach developed by Robert Fay, indicators are defined—two indicators for the missingness, one for module A and one for module B, column R and S (Fay, 1984).

In Figure 5-8, R and S are response indicators for A and B, respectively, where RS = 12 denotes table CA, RS = 21 denotes CB. Biemer uses log-linear path models to specify relationships among C, A, B, R, and S. The response mechanisms that are assumed depend on the model assumed for R and S—either an assumption that the response indicators R and S are related to the core (a missing at random assumption) or an assumption that they are related to some missing variable like A or B (a not missing at random, ignorable nonresponse). In this methodology, both ignorable and nonignorable response mechanisms can be estimated. He noted that matrix sampling, on the other hand, is primarily concerned with ignorable (missing at random) response mechanisms.

In matrix sampling, the data are missing completely at random, in which case R and S are related to anything except each other or, in worst case, missing at random where R and S might be related to some variable

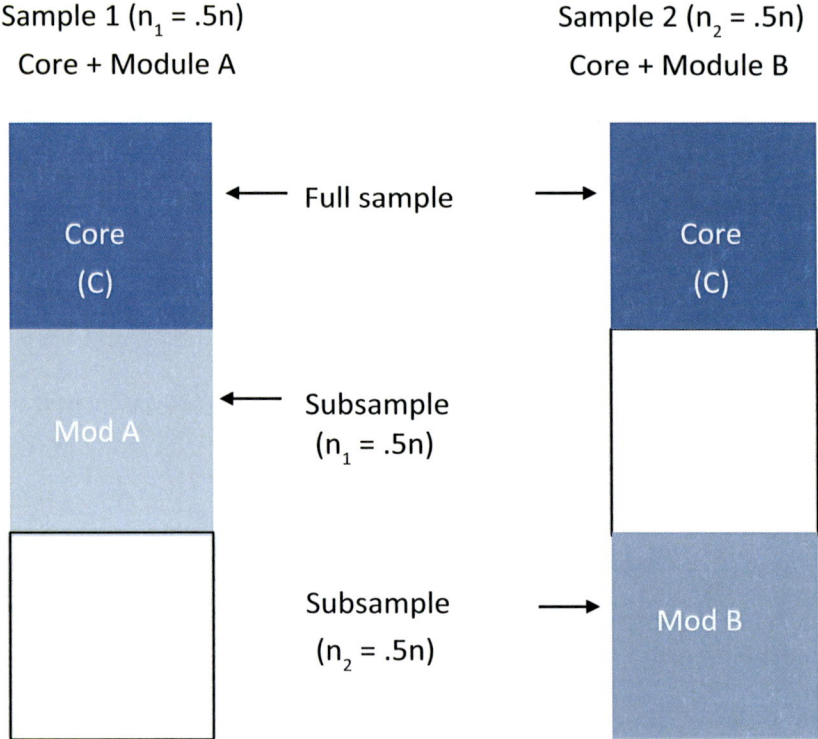

FIGURE 5-6 Simple matrix-sampling design.
SOURCE: Paul Biemer presentation at the Workshop on Respondent Burden in the American Community Survey, March 9, 2016. Available: http://sites.nationalacademies.org/cs/groups/dbassesite/documents/webpage/dbasse_173165.pdf [September 2016].

in the core because, depending upon a response in the core, the ACS might decide to ask some questions in module A or module B. He also asserted that precision of the model could be enhanced by involving other variables that might be correlated.

Biemer stated he is using FIML in analyzing the National Crime Victimization Survey—a panel survey with significant missingness. If complete case analysis were used, a huge number of cases would be discarded in the analysis. Based on this experience, he urged the Census Bureau to explore FIML, stating it is a viable approach for interval and model estimation and matrix sampling and is an alternative to multiple imputation methods. The standard errors are asymptotically equivalent for FIML and multiple

Subtable for C × A

	C = 1	C = 2
A = 1	$n_{a=1, c=1}$	$n_{a=1, c=2}$
A = 2	$n_{a=2, c=1}$	$n_{a=2, c=2}$

Subtable for C × B

	C = 1	C = 2
B = 1	$n_{b=1, c=1}$	$n_{b=1, c=2}$
B = 2	$n_{b=2, c=1}$	$n_{b=2, c=2}$

FIGURE 5-7 Notation for analyzing C, A, and B.
SOURCE: Paul Biemer presentation at the Workshop on Respondent Burden in the American Community Survey, March 9, 2016. Available: http://sites.nationalacademies.org/cs/groups/dbassesite/documents/webpage/dbasse_173165.pdf [September 2016].

imputation, and with FIML, it is possible to boost precision by bringing in the correlates, even those that are on separate modules. The downside, he said, is that the Census Bureau would need to write its own software or use specialized software like Mplus or Latent Gold. He noted that other software packages are now coming online with FIML procedures.

DISCUSSION

Colm O'Muircheartaigh (NORC at the University of Chicago) commented on an agenda for the next year or two—actions that are necessary for the ACS moving forward. The undertaking will be complicated, he said. If the goal is to produce estimates of one variable at a time, there is no need to intercorrelate variables, and various split questionnaire designs will work adequately in many cases. On the other hand, he said taking associations among all possible pairs requires more complicated designs. He referred to the experience of the redesign of the Add Health survey, for which he served

Incomplete data likelihood

$$\log L_{(\pi)} = \sum_{cab} n_{cab} \log \pi_{cab} \pi_{11|cab} + \sum_{ca} n_{ca} \log \sum_{b} \pi_{cab} \pi_{12|cab}$$
$$+ \sum_{cb} n_{abd} \log \sum_{a} \pi_{cab} \pi_{21|cab} + \sum_{c} n_{c} \log \sum_{ab} \pi_{cab} \pi_{22|cab}$$

where

$$\pi_{cab} = \Pr(C = c, A = a, B = a)$$
$$\pi_{rs|cab} = \Pr(R = r, S = s \mid C = c, A = a, B = a)$$

FIGURE 5-8 Likelihood assuming multinomial sampling.
SOURCE: Paul Biemer presentation at the Workshop on Respondent Burden in the American Community Survey, March 9, 2016. Available: http://sites.nationalacademies.org/cs/groups/dbassesite/documents/webpage/dbasse_173165.pdf [September 2016].

on the board of advisers. The field operation for that survey had become so large that it was impossible to collect all the data for all of the respondents. It was decided to drop some modules rather than split the sample, because Add Health is a longitudinal study and modularizing causes a loss of the longitudinal benefits and complicates analysis.

In order to appropriately design an approach to reducing burden, he urged the Census Bureau to consider a series of questions: What does the agency want to collect about the population over this period of time? Who are the clients? What are the statutory obligations? What are the user stakeholder obligations? What data should be collected? Is it appropriate to take an equal probability sample across the whole country? For example, designing for reliable data for every county, most of which have a very small fraction of the population, results in unnecessarily large samples in counties with larger populations. Is it appropriate to use the same sampling fraction for New York as for a rural county in Idaho?

He stated, in thinking about burden, the metric should be aggregate burden—the total effort caused to the sample by collecting the data. The largest savings in burden could come by reducing the number of respondents, so he urged a rethinking of the need for the size of sample by area. He suggested a "thought experiment" for the leadership at the Census Bureau to consider how to design an ACS by starting with the proposition that it needs to collect data about certain things for certain people for certain uses.

Heeringa commented on the presentations in this session. First, he agreed that the FIML method is a good exploratory tool, particularly for tabular data in which the inputs are categorized, even age distributions or groups. The method provides an indication of the ability to recover information or conversely the fraction of missing information from these sorts of planned missing data designs.

He also noted mass imputation, which was created by Donald Rubin to permit imputation of rectangularized datasets because software could not handle anything but complete data, would strike out cases with any missing data (Rubin, 1987). He observed that software today allows imputation directly in the analysis stream. If there is planned missingness, at least for the analytic uses within the census and by research users, he said he would provide the data in its missing data structure (and user guides) and then allow people to use the available software to construct their analysis. In making his recommendation, he noted that hierarchical consistency is desirable and in some cases necessary—block group totals should add to tract totals, which should add to county totals and then to state totals.

Biemer added that the National Survey of Drug Use and Health uses imputation for missing items, but the large number of items on the survey means not all are imputed. For those items that are not imputed, FIML is an alternative to prevent throwing out cases with missing data. It is possible to do mass imputation but leave open the option for FIML, he said.

Asiala stated that the sample design reflects, in part, the inherited practice from the decennial census long form for a sequence of adding more sampling rates over time. In 1990, there were three sampling rates, which grew to four sampling rates in 2000. The ACS has had five sampling rates, ranging from around 1.8 percent at the lowest up to 10 percent. As part of the 2011 reallocation of the ACS sample, the total number of sampling strata was increased to 16. The rates now vary from 0.5 percent per year for the largest tracts up to 15 percent for the smallest jurisdictions.

He reported consistency in data products is ensured by (1) a practice that, for all the tabulated data products, disclosure avoidance and any necessary data swapping occur prior to weighting the data, and (2) direct estimates are made from a weighted dataset. In this way, at any level of nested geography, by design there will be consistency all the way through.

Mathiowetz commented about matrix sampling and its relation to overall burden. One of the measures of burden is the length of the questionnaire and the number of contacts or interactions with the Census Bureau to process it. This measure would suggest that the people at the very tail end who have been asked to answer by Internet, by mail, by phone, and finally by face-to-face interview have the largest amount of burden. With matrix sampling and planned missingness, some of those people would have responded in total to the original full request. Perhaps rather than doing planned missingness, she suggested, the Census Bureau should consider a nonresponse questionnaire for those who are in the process toward the end. This practice would reduce the amount of information for those who have had multiple contacts and would link matrix sampling with a responsive design for nonresponse.

Heeringa and O'Muircheartaigh concurred that the idea is promising. The difference from the matrix sampling discussed earlier is that immediately a burden propensity is inserted into the process to produce an estimate. In addition, different sequences of methodology in terms of contacting respondents are worth considering. There are cost implications that should be considered as well, so this would be a major stakeholder issue for the ACS.

Brick observed that the literature on the length of the questionnaire suggests that only drastic changes would significantly affect burden. This is an important research question, he said. The experience at NORC, reported O'Muircheartaigh, has been that repeated phone calls seem to be what cause the most aggravation. Respondents do not mind as much if an interviewer shows up frequently because at least the survey is investing some effort.

A participant asked about how to identify where the burden is in the questionnaire. Understanding the source of the burden requires a focus on the ACS instrument itself. The participant opined that the instrument itself places a different burden on different people because it is not a static instrument. The most difficult questions are the ones repeated for every individual in the household, which raises the possibility of reducing burden by subsampling within the household and administering the difficult financial questions to just one person.

6

Tailoring Collection of Information from Group Quarters

It has been 10 years since a sample of group quarters (GQs)—correctional facilities, nursing homes, college dormitories, and the like—was added to the then-year-old American Community Survey (ACS), with the goal of more closely mirroring the design of the census long-form sample that the ACS was designed to replace. People in the group quarters sample units are now asked the same questions as household members about such topics as personal characteristics (e.g., disability, veteran status, and employment). The ACS housing questions are not asked, but information about type of facility is collected from the facility contacts.

The Census Bureau has found the collection, estimation, and analysis of GQ information to be quite challenging over this decade. For example, the small representation of group quarters in the monthly ACS samples has affected the quality of the estimates in many small areas that have large GQ populations relative to the total population.

In 2010, the Census Bureau asked the National Research Council (NRC), through a panel of the Committee on National Statistics, to review and evaluate the statistical methods used for measuring the GQ population. The panel's report contained several recommendations calling for improvements in the sample design, sample allocation, weighting, and estimation procedures and suggested further research to address the underlying question of the relative importance and costs of the GQ data collection in the context of the overall ACS (National Research Council, 2011).

Five years after that report and the introduction of the first ACS data products based on samples of both households and group quarters, the Census Bureau again requested the Committee on National Statistics to revisit issues surrounding the collection of data from group quarters as one of the key topics for this workshop. The steering committee developed an agenda that focused broadly on data collection methods for the group quarters component of the survey. The session featured an overview presentation by Judy Belton (U.S. Census Bureau) and additional presentations by Barbara Anderson (University of Michigan), Lauren Harris-Kojetin (National Center for Health Statistics), Andy Peytchev (University of Michigan), Michael Brick (Westat), and Colm O'Muircheartaigh (NORC at the University of Chicago). The session was chaired by steering committee member David Dolson (Statistics Canada), a member of the 2010 study panel.

THE FEASIBILITY OF TAILORING GROUP QUARTERS— SPECIFIC QUESTIONNAIRES IN THE ACS

Judy Belton addressed the feasibility of tailoring GQ-specific questionnaires in the ACS. She defined group quarters, provided background on ACS GQ data collection and the questionnaire items on the ACS form, and reported on the analysis and recommendations of an internal Census Bureau study to determine the feasibility of developing a GQ-specific questionnaire.

Group quarters, according to the official definition, are places where people live or stay in a group living arrangement that is owned and managed by an entity or organization that provides housing or services for its residents. Group quarters are divided into two groups—institutional and noninstitutional. Some examples of institutional are correctional facilities for adults, juvenile facilities, and nursing facilities. Some examples of noninstitutional GQs are college and university student housing, military barracks, and residential treatment centers.

The ACS samples about 18,000 GQs a year, classified as large (15 or more people living in the facility) and small (with fewer than 15 people). There are about 15,000 large GQs, mostly college and university student housing, nursing facilities, and correctional facilities, and nearly 3,000 small GQs, mostly group homes and workers' dorms.

The ACS collects data from a sample of about 1,600 facilities every month. From those facilities, the Census Bureau samples residents to participate in the ACS, with an average of about 10 people who participate in each GQ. The collection from 2006 through 2008 was paper-based, with field representatives using a paper questionnaire and conducting the interviews. In 2009, the ACS converted to collection using the computer-assisted personal interviewing (CAPI) instrument.

In every sample GQ, the interviewer first talks to a contact person—the gatekeeper—using an automated Group Quarters Facility Questionnaire (GQFQ). The GQFQ obtains the contact person's name, verifies the GQ type and address, and obtains the number of people who are living or staying at the GQ that day. Based on the number of people who are living or staying there on that day, the GQFQ randomly selects residents to participate in the survey. Field representatives attempt CAPI with the selected people. If CAPI cannot be conducted with a sample resident, the resident is asked to participate in a telephone interview. The next step is to conduct an interview with a proxy—a relative, parent, or contact at the facility. After failing to obtain an interview by these methods, the interviewer will leave a questionnaire with the sample resident and field staff will return to pick it up. If that fails, the Census staff member swears in a contact person at the facility, commissioning the contact person to drop the questionnaires off with the sample residents, and the field staff returns upon completion. As a last resort, field staff can use the facility's administrative records after obtaining permission from a Census Bureau regional office.

CAPI is the preferred method of data collection because it has skip patterns based on GQ type, Belton explained. For example, residents living in institutional GQs are not asked questions about how they travel to work or whether they have any children living in the facility. However, paper questionnaires are not totally eliminated because computers are not allowed and access is restricted in some facilities such as federal Bureau of Prisons facilities. For this type of group quarters, the forms are dropped off and mailed to the regional offices.

Belton stated that a special paper version of the CAPI questionnaire is provided to facility management and others as an example of the kinds of questions asked in the computer-based interview. The content of the GQ questionnaire is much like a housing unit questionnaire with only the social and economic questions. The GQ questionnaire does not contain plumbing and heating questions, but receipt of Supplemental Nutrition Assistance Program (SNAP) assistance, which is a housing question, is asked.

The Census Bureau is assessing the possibility of developing GQ-specific questionnaires. ACS data on response by type of institution and mode indicates that 84.3 percent of GQ interviews are conducted via CAPI while close to 16 percent are self-response using paper. For institutional GQ responses, only 7.6 percent are completed via paper questionnaires compared to 92.4 percent of CAPI responses. Reflecting the paper-based policies of the Bureau of Prisons, more paper is used in adult correctional—almost 12 percent—than in juvenile and nursing facilities.

Belton next reported on data on GQ paper responses by GQ type and respondent type. She reported that the majority of all GQ paper responses, almost 92 percent, were answered by the sample resident—80.4 percent of institutional paper GQ responses and almost 96 percent of noninstitutional paper responses were completed by the sample resident. Almost 50 percent of the responses came from proxies in nursing homes, and almost 39 percent were proxies in the other institutional GQ types. GQ CAPI responses varied from those who used paper. While the majority of CAPI interviews were completed by the sample resident, proxies completed a larger proportion of GQ CAPI responses than for paper responses—proxies accounted for 24 percent of responses for paper, and CAPI proxies accounted for almost 28 percent.

The Census staff also assessed missing data rates for GC reporters. The rates varied but were generally low regardless of the mode, except for the health insurance questions, for which there were reporting problems in both paper and CAPI. Comparing questionnaire missing data rates by institutional versus noninstitutional GQ types, the Census Bureau found that missing data rates were higher in institutional GQ types than in noninstitutional GQ types.

The institutions showing the highest missing data rates on the paper questionnaire were nursing facilities, probably related to the fact that health insurance is not available on administrative records. As might be expected, missing data rates were higher for paper responses completed by the proxies. Finally, the study found that facility-provided administrative records, while not used by the majority of GQ responders, were used by only 9 percent of GQ responses via paper questionnaire and 32 percent of GQ responses from CAPI interviews.

Belton summarized that the findings suggest that very few institutional GQ respondents self-respond using paper. Even responses to the eight questionnaire items proposed for removal in a paper version had relatively low missing data rates for these items, suggest to her that respondents were not burdened by the extra questionnaire items.

Developing and implementing a paper questionnaire for GQs would present operational issues related to the additional workload involved in assembling, distributing, and controlling paper questionnaires. She acknowledged that any changes in the processing and field-data collection procedures would need to be thoroughly tested.

Based on this review, the staff recommended that the Census Bureau should not create a GQ-specific questionnaire for institutional GQ types. The study also suggested the need to consider the option of offering the Internet option for students in college dorms, residents in military barracks, and perhaps some residents in group homes.

CENSUS SCIENTIFIC ADVISORY COMMITTEE FINDINGS

Barbara Anderson's comments drew from her work on the Census Scientific Advisory Committee Working Group on Group Quarters in the ACS. She, Robert Hummer (University of North Carolina), and Irma Elo

(University of Pennsylvania) constitute this working group. The committee developed several suggestions for consideration:

- Make the Internet version of the ACS available to noninstitutional GQ residents, especially in college dorms, military barracks, and group homes. Of the noninstitutional GQ respondents, 79 percent are college students or military personnel—very computer-savvy groups. Allowing noninstitutional GQ respondents to answer on the Internet should lower costs and improve data quality. Noninstitutional GQ respondents should be treated the same way as non-GQ respondents, which would substantially reduce the problems with obtaining data on GQ respondents, cutting them by about one-half.
- Ask only a short list of items to institutional GQ respondents that can be filled out by administrators, which would perhaps eliminate the paper form for institutional GQs. For institutional GQs, the committee suggested collecting data for a very short set of items—age, sex, race, ethnicity, and educational attainment. The advantage is that these data could be obtained completely from administrative sources. This option is attractive because the per-respondent cost of collecting GQ data is far higher than for non-GQ data, and it could eliminate the need for a paper form for institutional GQs.
- Flag imputed cases and values in the ACS Public Use Microdata Samples (PUMS). The ACS PUMS data should have an imputation flag for an entire imputed GQ respondent and also for specific imputed variables. As of now, there is no indication in the PUMS data whether particular variables, or even the whole case, were imputed from another source. Furthermore, if only a small set of items is collected for institutional GQ respondents, imputation strategies need to be rethought.
- Include more information in the ACS PUMS file on GQ type beyond the institutional/noninstitutional divide, currently the only breakdown available. A variable with a more detailed breakdown of GQ type in the ACS PUMS data would make these data more useful to users.

Anderson discussed the needs of the two main user communities for GQ data—municipalities, which are mainly concerned with average values and distributions, and researchers who want to build models and run multivariate models using the PUMS data. Her perception is that there has been overwhelmingly more concern for the needs of municipalities than PUMS data users. She pointed out the value of PUMS data analysis to the scholarly community, government, and policy.

NATIONAL NURSING HOME SURVEY EXPERIENCE

Lauren Harris-Kojetin said that she based her remarks on her experience at the National Center for Health Statistics (NCHS) Division of Healthcare Surveys, where she heads up the long-term care statistics program, obtaining data and producing national and state estimates on nursing facilities and residents. Nursing facilities are either the second or third largest set of group quarters in the ACS, she noted.

Harris-Kojetin stated that NCHS conducted the National Nursing Home Survey seven times between the 1970s and 2004. Similar to the ACS, the National Nursing Home Survey used in-person interviews to collect information on up to 10 sampled residents. Similar to the ACS protocol, National Nursing Home Survey field representatives went to a nursing facility at a scheduled interview appointment time and administered the questionnaire. Also similar to the ACS, while onsite, the field representatives then worked with the nursing facility respondent (typically the administrator) to sample 10 residents. In contrast to the ACS, the National Nursing Home Survey field representatives usually worked with the administrator or designated staff to complete the questionnaire for each of up to 10 sampled residents. Residents themselves were never interviewed, nor was there a need for a proxy, such as a relative.

Typically, the National Nursing Home Survey interviewers made only one visit to each reporting facility to complete all of the data collection both for the facility and the 10 sampled residents. However, in terms of burden, the interviewer may have spent several hours at the site. Harris-Kojetin posed a question about whether it is more burdensome to have one visit of several hours or shorter, multiple visits, with the nursing facility contact person having to coordinate with resident family members as proxies and meet with the field representative.

Prior to 2004, the National Nursing Home Survey used paper; CAPI was introduced in 2004. Starting in 2012, NCHS replaced the National Nursing Home Survey and its other existing long-term care provider surveys with the Biennial National Study of Long-Term Care Providers, which covered the supply, use, and characteristics of five major sectors of paid, regulated long-term care including nursing facilities. For the nursing facilities sector, NCHS now uses only administrative data from the Centers for Medicare & Medicaid Services (CMS).

The conversion to administrative data was made largely with the aim of lowering costs, but the effect has also been to reduce or eliminate burden for the nursing facilities. Harris-Kojetin suggested that the Census Bureau explore the feasibility of using administrative data maintained by CMS as an alternative to survey data collection for nursing facility group quarters in the ACS—specifically, the Minimum Data Set (MDS) 3.0.[1] Part of this exploration by the Census Bureau could include whether the actual MDS is needed versus a more processed, user-friendly version of the MDS data, such as the MDS Active Resident Episode Table, which NCHS uses.

Harris-Kojetin assessed that there is considerable overlap between the ACS Group Quarters questionnaire items and the MDS items. She pointed to overlap in demographic characteristics—name, gender, date of birth, race/ethnicity, and marital status—and in health insurance items like Medicare/Medicaid, language spoken, ability to hear or see, short-term and long-term memory issues, and ability to walk or in need of assistance with walking or dressing. Many of the GQ items collected in the ACS are not available from the MDS, she commented, so using MDS administrative data to completely replace survey data collection for nursing facilities in the ACS would require development of a much shorter version of the GQ set of items for nursing facilities.

There would be other benefits from using the administrative data from the MDS, she said. The universe of Medicare- and Medicaid-certified nursing facilities would be represented as well as the universe of residents in those nursing facilities, not just a sample as is done now in the ACS. By collecting from the universe rather than a sample, the ACS could address some of the GQ-related, small-area estimation issues in the ACS, at least for the variables that are comparable between administrative data sources and the ACS. Another benefit would be that using administrative data such as MDS would alleviate respondent burden on nursing facility staff, residents, and resident family proxies, she said.

Harris-Kojetin presented other scenarios for the ACS in addition to substantially shortening the ACS GQ item set for nursing facilities by using the MDS information and complete substitution for survey data collection. For instance, under the complete substitution scenario, there may be other administrative data sources, such as Department of Veterans Affairs administrative data for military service and service-connected disability and Social Security Administration information on work and income. The use of the other data sources would require that a unique identifier such as the Social Security number be available across administrative datasets and the ACS. Another scenario, rather than complete substitution for all sample nursing facility residents, would be to use administrative data sources after data collection for cases with specific survey items that have historically high missing rates.

---

[1] MDS 3.0 is an assessment done by nursing homes at regular intervals on every resident in a Medicare- or Medicaid-certified nursing home. It covers almost 99 percent of all nursing homes in the United States that are either Medicare- or Medicaid-certified. The MDS collects resident characteristics including demographic, functional, and clinical characteristics.

Harris-Kojetin commented on other aspects of the current ACS GQ design for nursing facilities. She stated that some of the alternative versions of the paper questionnaire that are being considered to enhance the ACS GQ design appear less relevant to nursing facilities. In the 2014 ACS, 99 percent of nursing facility responses were submitted through CAPI, so she concluded that it would not be worthwhile to create another questionnaire and have to deal with the logistics and the costs of an additional paper form for 1 percent of nursing facilities. She also addressed the use of proxy respondents for the resident questionnaire response process in nursing facilities. According to the 2014 National Study of Long-Term Care Providers, half of nursing home residents have Alzheimer's or other dementias, a likely cause of the fact that proxies complete about one-half of ACS nursing facility resident questionnaires. It would be useful to gain further understanding of the ACS Group Quarter Resident Questionnaire completion process in nursing facilities regarding the quality of data under the three main scenarios: where the resident self-completes, where the questionnaire is completed by a proxy who is a relative of the selected respondent, or where he survey is completed by a proxy who is a nursing facility staff.

Finally, she discussed potential uses of the CMS's Nursing Home Compare Website (data.medicare.gov)—a publicly available Website that lists all Medicare- and Medicaid-certified nursing homes; provides the name, address, phone number, bed size, current number of residents, and number of certified beds; and the location of nursing facilities. She suggested Nursing Home Compare as a valuable resource for updating the nursing facility information on the master address file between decennial census years. Further, the Website provides for downloading the file, including the federal provider number, for each nursing facility.

In summary, she pointed out that NCHS has used administrative data in two ways: before 2012, to substitute for personal collection from selected respondents by using records maintained by the nursing facility, and starting in 2012, to avoid going to a nursing facility entirely by using administrative data from another federal agency.

ACS CONSTRAINTS RELATED TO GROUP QUARTERS

Andy Peytchev presented and discussed a list of constraints under which the ACS operates that influence the environment for collecting, processing, estimating, analyzing, and publishing data on group quarters. A key constraint is the need to continue to collect the same data from the GQ questionnaires that are currently collected. He commented that reducing burden will involve design changes, and every design change involves tradeoffs. In the ACS, the tradeoffs are complex because of the multiple components of the program. He offered a conceptualization of the relationship between burden, variance, bias, cost, and other quality dimensions.

The relationships are complex. For example, within bias and variance, there are nonresponse, coverage, and measurement effects. All these sources of error need to be measured and balanced against other changes in the survey, because any time an intervention or a change in protocol is made to affect burden, at least some of the other components will be affected.

He said the Census Bureau could use two different paper forms to minimize the unnecessary questions; another alternative would be to embed additional skip logic. He stated that the option of dropping the paper and pencil interviewing (PAPI) instrument altogether seemed like a radical choice, but a more feasible alternative would be to implement a Web option for most of the PAPI. Alternatively, the paper instrument could be employed in an even more limited and targeted manner. Targeting would require understanding what burden means, to whom the burden accrues, and, once the data are collected, from whom the suspect information (in terms of bias and variance) is obtained.

Peytchev noted that the proposed paper instrument labeled 48 items, and, counting the subquestions, it totals about 80 questions. The proposed GQ survey would be shorter than the household instrument by half, and the items appear to be simpler to answer. The proposed paper survey, in terms of Norman Bradburn's framework, is simpler in at least three of the four dimensions of burden. However, in some facilities, one person would have to answer for everybody else and, when that happens, the individual burden could explode to potentially 800 or 1,600 questions.

Based on his assessment, he suggested the following:

- Consider limiting the use of the paper instrument, whether only to self-administration, to specific types of facilities, or some combination of the two, as a short-term solution to this aspect of burden.
- Consider reevaluating sampling for some GQ types, for example, increase the facility sampling rates and decrease the within-facility sampling rates to reduce burden. (Currently, the sampling rate within selected facilities with 10 or fewer residents is 100%.)
- Evaluate the impact on the survey estimates. It is important to be cognizant of the implications of burden on the properties of the survey data that are being collected.

ADDITIONAL PERSPECTIVES ON GROUP QUARTERS

Michael Brick summarized his perception of some of the key suggestions related to GQ:

- Split off the institutional from the noninstitutional GQs with a much smaller set of questions relevant to each. The remaining questions could be completed by administrative records.
- For institutional GQs, eliminate the CAPI interview and the paper instrument, and import administrative records or, where administrative data are not available, use CAPI.

On the noninstitutional side, he supported the Internet option particularly for college student facilities and military barracks. He further suggested that the Census Bureau:

- Limit the number of times that field representatives go back to the same facility over the year.
- Ensure the questionnaire makes sense to the intended respondents.
- Address the issue of possible double burden for college students, that is, the burden that comes when parents within their household have a student living in a dorm and the student's information is also collected in the GQ survey.

Colm O'Muircheartaigh praised the Census Bureau and characterized the ACS as a triumph, given where it started and what it has become. He advocated eliminating the term "group quarters," which he called "an almost meaningless term." To him, the term has led to conflicts in dealing with an extraordinarily heterogeneous collection of arrangements under one heading. Instead, he suggested the principle of stratification, which is one of the key means by which the survey can treat different parts of the population differently. It is important, he said, to think about the challenges of data collection, not about labeling entities under one term that has nothing to do with how they should be approached.

DISCUSSION

Connie Citro commented on the history of the treatment of group quarters in the ACS. Most household surveys cover only the noninstitutional population because the institutional population is only 3 percent of the U.S. population, has been pretty steady over the past few decades, and is hard to count. The census mandate to collect information about everybody and not just the civilian, noninstitutionalized population was carried over to the ACS. The 2010 NRC panel considered the appropriateness of the GQ survey but determined users want the information. To some extent, she noted, users come to this conclusion failing to understand that the ACS does not provide detailed data about group quarters. It provides some state totals by type of GQ but no detail for the GQ population on characteristics such as education, health insurance, and disabilities. Furthermore, for small areas, the information can distort characteristics because the GQ population is very different from the rest of the population in an area.

A participant asked about the treatment of group quarters in coverage measurement. It is important to recognize that unrelated people in group quarters should mostly be treated separately. Dolson replied that a count of the number of unrelated people in an area is important, but the concept is not clean. For example, universities or private apartment complexes offer individual leases for group quarters, making count of unrelated individuals at an address unreliable. Belton agreed that this is an important issue and reported that the Census Bureau is working on a definition that encompasses these new arrangements. O'Muircheartaigh suggested people in these arrangements should be classified as living in apartments. Brick concurred, adding that people in these arrangements are part of the noninstitutionalized population and, as such, should be given the noninstitutional questionnaire to complete.

Belton added that assisted living facilities raise some of the same issues. They are classified as housing units, but some have a floor or a wing with continuous skilled nursing care. According to the ACS definition, they are group quarters.

A participant asked how the Census Bureau would conduct sampling at GQs if, as has NCHS, administrative records become the primary means of obtaining information from nursing facilities and field representatives no longer physically go to the locations. Harris-Kojetin responded that when NCHS transitioned to using only administrative data from CMS for the nursing home sector, sampling was no longer required. The universe of residents at the Medicare- and Medicaid-certified nursing facilities was obtained from administrative records. New issues did present themselves, however, such as the reference period and how often the information is updated. For example, the MDS information on the Website is updated quarterly. The GQ approach for the ACS has a separate sample every month.

Salvo asked about the classification of multiple-use GQs. With the aging of the population and the complexity of some assisted living arrangements, step-up arrangements are becoming more popular where part of the facility is a nursing home, part assisted living, and part independent apartments, he observed. These arrangements are difficult to disentangle from an address standpoint. It may mean the Census Bureau has to internally collect the data and then decide how to categorize them, he suggested.

O'Muircheartaigh added that the issue of estimation was one of the debates when the ACS was introduced. The position in the Census Bureau and others in the demographic community was that the ACS could not be matched with the census because the ACS collected these data on an ongoing basis throughout the year and the census was clearly defined as being on the first of April. This would lead to confusion. However, he advocated making decisions on the basis of the information that is to be produced or estimated and then to collect data to make it possible to estimate that information.

7

Future Directions

The workshop's final session, chaired by Joseph Salvo, focused on some of the ideas discussed at the workshop in regard to the four topics: (1) use of matrix sampling to reduce actual and, potentially, perceived burden; (2) direct substitution of administrative records to replace survey questions to reduce actual and, potentially, perceived burden; (3) communication and mail-package messaging to reduce perceived burden and encourage response by Internet or mail; and (4) tailoring and reduction of questionnaires for residents of institutionalized and noninstitutionalized group quarters (GQ). He set the framework for the discussion by pointing out that the American Community Survey (ACS) is 10 years into implementation, with countless changes over its short history. Changes have been made in the addition of group quarters, editing of the GQ population, weighting procedures that have been employed, nonresponse follow-up, sample allocation, and many other facets of the survey.

Today, according to Salvo, the ACS is experiencing an existential crisis. The pressure to make the ACS voluntary, budget threats, and other appropriation issues cause a great deal of consternation. In this environment, he asserted, it is the responsibility of the Census Bureau and its stakeholders to achieve the right balance between burden and the collection of data, while managing costs. The Census Bureau needs input from its stakeholders in order to meet these challenges, he said.

He called for suggestions on future directions for the work on reducing the ACS burden. Several participants offered suggestions, which are summarized below.

- Develop a communication program to distinguish between the 2020 Census and the ACS. A participant added that it is important to the ACS that the Census Bureau will soon be launching an integrated partnership and communication program for the 2020 census. The ACS will continue to be in the field through 2020, and there will have to be communication about the difference between the ACS and the decennial census because both of them will be collected at the same time. She proposed that, rather than just waiting to figure out what should be said in 2020, the communication program should now be thinking about how to distinguish and brand the ACS and the decennial.
- Obtain data user feedback. The participant's second proposal for the future was that the Census Bureau should obtain data user input and feedback when matrix sampling, multiphased sampling, or other major changes are developed, tested, and introduced. She projected potentially substantial changes in the usability of the ACS when these changes occur.
- Update dissemination tools. The participant's third proposal was that as the ACS products change in terms of sampling and structure, the dissemination tools need to change. There needs to be communication between the Census Bureau's Center for Enterprise Dissemination Services and Consumer Innovation (CEDSCI), which is developing new dissemination tools and platforms, and the people researching changes in the ACS, she said. The CEDSCI platforms that are being developed need to be flexible enough to accommodate these kinds of changes.
- Approach integration of administrative data with caution. Connie Citro concurred that the Census Bureau needs to be cautious about commercial and other outside data that might be brought into the ACS products, as they may not be of very good quality.
- Repurpose the program. If and when the outside data are integrated, Citro proposed renaming the ACS as the American Community Information Program. The goal would be to provide the small-area multivariate information that the country has long expected and gotten, previously every 10 years from the long form and now in the continuous ACS. She envisioned an American Community Information Program enriched with people's Social Security income and the Supplemental Nutrition Assistance Program data, for example. The program could be cooperative in that the Census Bureau would exert review and oversight and enforce quality standards for a lot of the housing and other data. She proposed that the Census Bureau role would be to improve data quality where possible and to reduce burden where it could.

- Link to geographic tools. Citro further suggested that Google Street View would be great for adding indicators of the type of housing and other location factors, such as proximity to services, to create a valuable American Community Information Program.
- Conduct research on integrating administrative and survey data. Another participant supported research and development leading to optimal integration of administrative data and survey data. The survey data can inform with known properties about the population and distributions and characteristics that must be available on an ongoing basis for working and modeling administrative data.
- Sharpen Census Bureau branding. Another participant commented on future directions in the ACS branding. There is confusion with the name of the organization, the Census Bureau. Consequently, newspapers reporting on the ACS say the data are from the U.S. Census. There is a need to figure out a way to distinguish the Census Bureau from just the U.S. Census, the participant said.
- Conduct further dialogue with administrative data producers. On the topic of the ACS communication, a participant suggested that the Census Bureau open additional dialogue with agencies that produce administrative data in order to learn more about programs that generate data and the potential benefits of linking the data. Amy O'Hara responded that Census Bureau staff interacts with the program agencies. Some agencies do not want to share their data with the Census Bureau for statistical or research uses unless they see a benefit. The exception is the Internal Revenue Service, which, by statute, is directed to share data with the Census Bureau. Nonetheless, the Census Bureau has an ongoing dialogue with the Social Security Administration, Department of Housing and Urban Development, Department of Veterans Affairs, and Department of Agriculture, and within Agriculture, with both the Food and Nutrition Service and the Economic Research Service, she said. These contacts determine what data exist and where there may be mutual benefits from sharing data that the Census Bureau helps to improve.
- Strengthen interdisciplinary development. A participant supported the idea of including more interdisciplinary staff from the Census Bureau or other federal agencies in the ACS developmental efforts. The participant suggested a team of statisticians and computer scientists to increase the value of the enterprise.
- Follow up on topics from this workshop in expert meetings. Amy O'Hara suggested some topics for the upcoming expert meetings on ACS burden reduction. First, she said, is the need to better understand the error structure—more work needs to be done to

measure the various dimensions of quality. The second is to move into more model-based, hybrid estimates along the lines of the work on Medicaid and SNAP underreporting with Current Population Survey data. Her third suggestion was to engage in at least one pilot that would sharpen income measurement and help overcome the challenges faced with the income question as written and the income sources that the Census Bureau can access.

- Conduct research and a methods test for improved definition of burden. Paul Biemer suggested work on the definition of burden and how to quantify it. A better definition would include the perception of the respondent on burden. A good definition will enable a measurement of that burden as a baseline so progress toward reducing the burden over time can be measured. It would enable setting a goal, such as over some period of time reducing the number by some percent. He also supported development of a methods test panel for experimenting with reducing the ACS burden.

At the conclusion of the session, Salvo thanked the participants on behalf of the steering committee and invited the participants to share any additional ideas with the workshop organizers.

References

Bradburn, N. (1978). Respondent burden. *Proceedings of the Survey Research Methods Section of the American Statistical Association, 1978*, 35-40. Available: https://www.amstat.org/sections/srms/proceedings/papers/1978_007.pdf [September 2016].

Chetty, R. (2012). *Time Trends in the Use of Administrative Date for Empirical Research.* National Bureau of Economic Research Summer Institute. Available: http://www.rajchetty.com/chettyfiles/admin_data_trends.pdf [September 2016].

Chipperfield, J., and Steel, D.G. (2012). Multivariate random effect models with complete and incomplete data. *Journal of Multivariate Analysis, 109*, 146-155.

Chipperfield, J., Barr, M., and Steel, D. (2013). *Split Questionnaire Designs: Are They an Efficient Design Choice?* Presented at the 59th World Statistics Congress of the International Statistical Institute, August 25-30, Hong Kong. Available: http://www.statistics.gov.hk/wsc/IPS033-P1-S.pdf [September 2016].

Cialdini, R.B. (1988). *Influence: Science and Practice.* Glenview, IL: Scott Foresman.

Commission on Federal Paperwork. (1977). *A Report of the Commission on Federal Paperwork: Final Report Summary.* Washington, DC. Available: https://babel.hathitrust.org/cgi/pt?id=pur1.32754075989214;view=1up;seq=3 [September 2016].

Davern, M., Meyer, B.D., and Mittag, N. (2015). *Creating Improved Survey Data Products Using Linked Administrative-Survey Data.* Paper presented at the Annual Meeting of the Federal Committee on Statistical Methodology, December 8-9, Washington, DC. Available: http://fcsm.sites.usa.gov/reports/research/2015-research/ [September 2016].

Dillman, D.A. (2000). *Mail and Internet Surveys: The Tailored Design Method* (2nd ed.). New York: John Wiley.

Dillman, D.A., Smyth, J.D., and Christian, L.M. (2014). *Internet, Phone, Mail and Mixed-Mode Surveys; The Tailored Design Method* (4th ed.). Hoboken, NJ: John Wiley.

Fay, R.E. (1984). Replication approaches to the log-linear analysis of data from complex surveys. *Recent Developments in the Analysis of Large-Scale Data Sets* (pp. 95-118). Proceedings of a seminar held in Luxembourg, November 16-18, 1983. Luxembourg: Office for Official Publications of the European Communities. Available: file:///C:/Users/ywise/Downloads/CAAB84006ENC_001%20(1).pdf [October 2016].

Frankel, J. (1980). *Measurement of Respondent Burden: Study Design and Early Findings.* Technical Report. Washington, DC: Bureau of Social Science Research.

Fricker, S., Gonzalez, J., and Tan, L. (2011). *Are You Burdened? Let's Find Out.* Paper presented at the Annual Conference of the American Association for Public Opinion Research, October, Phoenix, AZ.

Fricker, S., Kreisler, C., and Tan, L. (2012). *An Exploration of the Application of the PLS Path Modeling Approach to Creating a Summary Index of Respondent Burden.* Paper presented at the Joint Statistical Meeting, August, San Diego, CA.

Galullo, D. (2013). Everything you know about branding is wrong. *Forbes,* December 3. Available: http://www.forbes.com/sites/onmarketing/2013/12/03/everything-you-know-about-branding-is-wrong/#448f1aa66685 [September 2016].

Giesen, D. (2012). *Exploring Causes and Effects of Perceived Response Burden.* Paper presented at the International Conference on Establishment Surveys. Available: https://ww2.amstat.org/meetings/ices/2012/papers/302171.pdf [October 2016].

Gonzalez, J.M., and Eltinge, J.L. (2007). Multiple matrix sampling: A review. *American Statistical Association: Proceedings of the Section on Survey Research Methods.* Available: http://ww2.amstag.org/sections/srms/Proceedings [November 2016].

Gonzalez, J.M., and Eltinge, J.L. (2010). Optimal survey design: A review. *American Statistical Association: Proceedings of the Section on Survey Research Methods.* Available: http://ww2.amstag.org/sections/srms/Proceedings [November 2016].

Groves, R.M., and Couper, M. (1998). *Nonresponse in Household Interview Surveys.* New York: John Wiley.

Groves, R.M., and Heeringa, S. (2006). Responsive design for household surveys: Tools for actively controlling survey errors and costs. *Journal of the Royal Statistical Society Series A: Statistics in Society, 169*(Pt. 3), 439-457.

Groves R.M., Singer E., and Corning, A.D. (2000). A leverage-saliency theory of survey participation: Description and illustration. *Public Opinion Quarterly, 64,* 299-308.

Hoogendoorn, A.W. (2004). A questionnaire design for dependent interviewing that addresses the problem of cognitive satisficing. *Journal of Official Statistics, 20,* 219-232.

Hoogendoorn, A.W., and Sikke, D. (1998). Response burden and panel attrition. *Journal of Official Statistics, 14*(2), 189-205.

Martin, E., Abreu, D., and Winters, F. (2001). Money and motive: Effects of incentives on panel attrition in the survey of income and program participation. *Journal of Official Statistics, 17,* 267-284.

Merkouris, T. (2015). An efficient estimation method for matrix survey sampling. *Survey Methodology, 41*(1), 237-262.

Meyer, B.D., Mok, W.K.C., and Sullivan, J.X. (2015). *Household Surveys in Crisis.* NBER Working Paper No. 21399. Cambridge, MA: National Bureau of Economic Research. Available: http://www.nber.org/papers/w21399.pdf [September 2016].

National Research Council. (1991). *Improving Information for Social Policy Decisions: The Uses of Microsimulation Modeling. Volume 1. Review and Recommendations.* C.F. Citro and E.A. Hanushek (Eds.). Panel to Evaluate Microsimulation Models for Social Welfare Programs. Committee on National Statistics, Commission on Behavioral and Social Sciences and Education. Washington, DC: National Academy Press.

National Research Council. (2011). *Measuring the Group Quarters Population in the American Community Survey: Interim Report.* K. Marton and P.R. Voss (Eds.). Panel on Statistical Methods for Measuring the Group Quarters Population in the American Community Survey. Committee on National Statistics, Division of Behavioral and Social Sciences and Education. Washington, DC: The National Academies Press.

National Research Council. (2013a). *Benefits, Burdens, and Prospects of the American Community Survey: Summary of a Workshop.* D.L. Cork, Rapporteur. Committee on National Statistics, Division of Behavioral and Social Sciences and Education. Washington, DC: The National Academies Press.

National Research Council. (2013b). *Nonresponse in Social Science Surveys.* Panel on a Research Agenda for the Future of Social Science Data Collection. Committee on National Statistics, Division of Behavioral and Social Sciences and Education. Washington, DC: The National Academies Press.

National Research Council. (2015). *Realizing the Potential of the American Community Survey: Challenges, Tradeoffs, and Opportunities.* Panel on Addressing Priority Technical Issues for the Next Decade of the American Community Survey. Committee on National Statistics, Division of Behavioral and Social Sciences and Education. Washington, DC: The National Academies Press.

Navarro, A., and Griffin, R.A. (1993). Matrix sampling designs for the year 2000 census. *American Statistical Association: Proceedings of the Section on Survey Research Methods.* Available: http://ww2.amstag.org/sections/srms/Proceedings [November 2016].

President's Council on Physical Fitness and Sports. (1985). *National School Population Fitness Survey.* Available: http://eric.ed.gov/?id=ED291714 [September 2016].

Raghunathan, T.E., and Grizzle, J.E. (1995). A split questionnaire survey design. *Journal of the American Statistical Association, 90,* 54-63.

Rubin, D.B. (1987). *Multiple Imputation for Nonresponse in Surveys.* New York: John Wiley.

Sharp, L.M., and Frankel, J. (1983). Respondent burden: A test of some common assumptions. *Public Opinion Quarterly, 47,* 36-53.

Thomas, N., Raghunathan, T.E., Schenker, N., Katzoff, M.J., and Johnson, C.L. (2006). An evaluation of matrix sampling methods using data from the National Health and Nutrition Examination Survey. *Survey Methodology, 32,* 217-232.

U.S. Census Bureau. (2014). *American Community Survey Content Review Summit.* Available: http://www2.census.gov/programs-surveys/acs/operations_admin/2014_content_review/ACSContentReviewSummit.pdf [September 2016].

U.S. Census Bureau. (2015a). *Agility in Action: A Snapshot of Enhancements to the American Community Survey.* Washington, DC: American Community Survey Office, U.S. Census Bureau. Available: http://www.census.gov/content/dam/Census/programs-surveys/acs/operations-and-administration/2015-16-survey-enhancements/Agility_in_Action.pdf [November 2016].

U.S. Census Bureau. (2015b). *Reducing Respondent Burden in the American Community Survey: A Feasibility Assessment of Methods to Ask Survey Questions Less Frequently or of Fewer Respondents.* Washington, DC: American Community Survey Office, U.S. Census Bureau. Available: https://www.census.gov/content/dam/Census/programs-surveys/acs/operations-and-administration/2015-16-survey-enhancements/Reducing_Burden_ACS_Feasibility_Assessment.pdf [September 2016].

U.S. Department of Commerce. (2015). Submission for OMB review; comment request. *Federal Register, 80*(103), 30655-30659.

U.S. Office of Management and Budget. (2006). *Standards and Guidelines for Statistical Surveys.* Washington, DC: U.S. Office of Management and Budget. Available: https://www.whitehouse.gov/sites/default/files/omb/inforeg/statpolicy/standards_stat_surveys.pdf [September 2016].

Zelenak, M.F., and David, M.C. (2013). *Impact of Multiple Contacts by Computer-Assisted Telephone Interview and Computer-Assisted Personal Interview on Final Interview Outcome in the American Community Survey.* Washington, DC: Decennial Statistical Studies Division, U.S. Census Bureau. Available: https://www.census.gov/content/dam/Census/library/working-papers/2013/acs/2013_Zelenak_01.pdf [September 2016].

Appendix A

Workshop Agenda

WORKSHOP ON RESPONDENT BURDEN IN THE
AMERICAN COMMMUNITY SURVEY
March 8-9, 2016
The National Academies of Sciences, Engineering, and Medicine
Keck Center, 500 Fifth Street, NW, Room 100
Washington, DC

TUESDAY, MARCH 8

8:00 am   *Continental Breakfast Available*

8:30 am   Welcome, Introductions
Connie Citro, Committee on National Statistics
John Thompson, Census Bureau

8:40 am–
9:30 am   Session I: Understanding Respondent Burden in the American Community Survey
Moderator: Brian Harris-Kojetin, Committee on National Statistics

- *Framework and Approach to Respondent Burden in This Workshop*
  Joe Salvo, New York City Department of City Planning

- *ACS Successes and Challenges Regarding Respondent Burden*
  Deborah Stempowski, Census Bureau
- *Defining, Measuring, and Mitigating Respondent Burden*
  Scott Fricker, Bureau of Labor Statistics

9:30 am–
11:00 am   Session II: Communicating with Respondents: Materials and the Sequencing of Those Materials
Moderator: Linda Gage, State of California (retired)

- *American Community Survey Mail Contact Strategy and Research*
  Elizabeth Poehler, Census Bureau
- *Improving Response to the American Community Survey (ACS)*
  Don Dillman, Washington State University
- *ACS Respondent Materials and Sequencing: Application of a Responsive and Adaptive Survey Design Framework*
  Andy Peytchev, University of Michigan
- *Communicating with Respondents: Material and Sequencing in the ACS*
  Nancy Mathiowetz, University of Wisconsin–Milwaukee (emerita)
- *Communicating the American Community Survey's Value to Respondents*
  Andrew Reamer, George Washington University

11:00 am–
11:15 am   *Break*

11:15 am–
12:30 pm   Session III: The American Community Survey: Communicating the Importance to the American Public
Moderator: Nancy Mathiowetz, University of Wisconsin–Milwaukee (emerita)

Discussion with:
- Sandra Bauman, Bauman Research & Consulting
- Betty Lo, Nielsen
- George Terhanian, NPD Group

12:30 pm–1:30 pm	*Lunch provided for all attendees*
1:30 pm–3:00 pm	Session IV: Tailoring the Group Quarters Questionnaire Moderator: David Dolson, Statistics Canada *The Feasibility of Tailoring Group Quarters Specific Questionnaires in the American Community Survey* Judy Belton, Census Bureau Discussion with: • Barbara Anderson, University of Michigan • Mike Brick, Westat • Lauren Harris-Kojetin, National Center for Health Statistics • Colm O'Muircheartaigh, NORC at the University of Chicago • Andy Peytchev, University of Michigan
3:00 pm–3:15 pm	*Break*
3:15 pm–4:45 pm	Session V: Use of Administrative Records to Reduce Burden and Improve Quality Moderator: Julia Lane, New York University • *Use of Administrative Records to Reduce Burden and Improve Quality* Amy O'Hara, Census Bureau • *Comments on "Use of Administrative Records to Reduce Burden and Improve Quality"* Paul Biemer, RTI International • *Use of Administrative Records to Reduce Burden and Improve Quality: A Discussion* Mike Davern, NORC at the University of Chicago
4:45 pm–5:15 pm	Session VI: Wrap-up: Discussion of Key Issues Moderator: Linda Gage, State of California (retired)
5:15 pm	*Adjourn*

WEDNESDAY, MARCH 9

8:30 am *Continental Breakfast Available*

9:00 am–
10:30 am Session VII: Matrix Sampling and Multiphase Sampling
Moderator: Dave Hubble, Westat

- *Overview of Feasibility Assessment of Using Matrix Sampling and Other Methods to Reduce Respondent Burden*
  Mark Asiala, Census Bureau
- *Utilizing Matrix Sampling to Reduce Respondent Burden*
  Jeffrey Gonzalez, Bureau of Labor Statistics
- *Planned "Missingness" Designs and the ACS*
  Steve Heeringa, University of Michigan

Additional discussion by:
- Paul Biemer, RTI International
- Mike Brick, Westat
- Colm O'Muircheartaigh, NORC at the University of Chicago

10:30 am–
10:45 am *Break*

10:45 am–
12:00 pm Session VIII: Modeling and Imputation
Moderator: John Eltinge, Bureau of Labor Statistics

- *Modeling and Imputation Discussion*
  Mike Brick, Westat
- *A Full Information Maximum Likelihood (FIML) Approach to Compensating for Missing Data in Matrix Sampling*
  Paul Biemer, RTI International

Additional discussion by:
- Jeffrey Gonzalez, Bureau of Labor Statistics
- Steve Heeringa, University of Michigan
- Colm O'Muircheartaigh, NORC at the University of Chicago

APPENDIX A

12:00 pm–
1:00 pm   *Lunch provided for all attendees*

1:00 pm–
2:15 pm   **Session IX: Administrative Records and the ACS: Future Directions**
Moderator: Linda Gage, State of California (retired)

        Discussion with:
- Frauke Kreuter, University of Maryland
- Julia Lane, New York University

2:15 pm–
3:00 pm   **Session X: Wrap-up: Discussion of Key Issues**
Moderator: Joe Salvo, New York City Department of City Planning

3:00 pm   *Adjourn*

Appendix B

Biographical Sketches of Steering Committee and Presenters

Barbara A. Anderson (*Presenter*) is the Ronald Freedman collegiate professor of sociology and population studies at the University of Michigan, where she also has been director of the Population Studies Center and the Center for Russian and East European Studies. She was a Guggenheim fellow and has been a member of the Institute for Advanced Study and the Center for Advanced Study in the Behavioral Sciences. She has published widely on issues of demographic methods, data quality, and population and development in the former Soviet Union, China, and South Africa. She has consulted with Statistics South Africa, Statistics Estonia, the China National Bureau of Statistics, the Turkish Statistical Institute, and the U.S. Census Bureau. She is currently chair of the U.S. Census Bureau's Scientific Advisory Committee. She received her bachelor's in mathematics from the University of Chicago and Ph.D. in sociology from Princeton University.

Mark Asiala (*Presenter*) is chief of the American Community Survey (ACS) Statistical Design area in the Decennial Statistical Studies Division of the U.S. Census Bureau. He started working at the Census Bureau in 1999 on the Census 2000 Accuracy and Coverage Evaluation. Since 2002, he has worked in the statistical design area of the ACS, particularly in the areas of estimation and disclosure avoidance. He has also been a member of the Census Bureau's Disclosure Review Board since 2008. He received a bachelor's in mathematics from the University of Michigan and an M.S. in mathematics from Georgia State University.

Sandra L. Bauman (*Presenter*) is founder and principal of Bauman Research & Consulting, LLC, a boutique research company. She has designed and managed hundreds of studies for corporate and nonprofit clients in the areas of branding, positioning, corporate image, messaging, strategic marketing, and customer satisfaction and loyalty. She is an expert in quantitative methodologies, including telephone, Internet, and mail surveys. She is also a trained and experienced focus group moderator and facilitator. She is a long-time member of the American Association for Public Opinion Research and currently serves on its executive council. She is an active member of the Marketing Research Association (MRA) and holds MRA's professional researcher certification at the expert level. She holds a B.A. in journalism from Drake University and an M.S.J. and a Ph.D. in communication research from Northwestern University.

Judy G. Belton (*Presenter*) is chief of the Group Quarters Data Collection Branch in the American Community Survey Office (ACSO). She has been with the U.S. Census Bureau for 28 years, including 9 in the ACSO. She has worked on several other surveys and the decennial census. She created and leads the Census Bureau's Group Quarters Working Group, which makes recommendations and/or decisions about group quarter (GQ) data collection methodologies with the goal of improving GQ data collection across the Census Bureau.

Paul Biemer (*Presenter*) is a distinguished fellow in statistics at RTI International and associate director of survey research and development in the Odum Institute at the University of North Carolina. He has more than 35 years of experience in survey methodology, complex survey design, and data analysis and has written more than 100 publications related to these areas. He is a fellow of the American Statistical Association and the American Association for the Advancement of Science and an elected member of the International Statistical Institute. He holds a number of awards for his contributions to the field of survey methodology and statistics, including the Morris Hansen Award. He holds a B.S. in mathematics and an M.S. and a Ph.D. in statistics from Texas A&M University.

J. Michael Brick (*Presenter*) is a vice president at Westat, where he is co-director of the Survey Methods Unit and associate director of the statistical staff. He also is a research professor in the Joint Program in Survey Methodology at the University of Maryland. He has more than 40 years of experience in survey research and is a fellow of the American Statistical Association and an elected member of the International Statistical Institute. He holds a B.S. in mathematics from the University of Dayton and an M.S. and a Ph.D. in statistics from the American University.

Michael Davern (*Presenter*) is an executive vice president of research and director of health care research at NORC at the University of Chicago. In this role he oversees NORC's three health departments and serves as the department head for health care research. The departments conduct survey research and analytic research, as well as provide technical assistance to clients that include the federal government, foundations, and commercial enterprises. Davern also has expertise in survey research, health data, data linkage, U.S. Census Bureau data, and the use of these data for policy research simulation and evaluation. He holds a B.A. in sociology from St. John's University, an M.A. in sociology from Colorado State University, and a Ph.D. in sociology from the University of Notre Dame.

Donald A. Dillman (*Presenter*) is Regents professor in the Department of Sociology at Washington State University (WSU). He also serves as the deputy director for research and development in WSU's Social and Economic Sciences Research Center. He maintains an active research program on the improvement of survey methods and how information technologies influence rural development. From 1991 to 1995, he served as the senior survey methodologist in the office of the director at the U.S. Census Bureau. He has served as investigator on more than 80 grants and contracts and written 13 books and more than 235 other publications. He is a fellow of the American Association for the Advancement of Science and the American Statistical Association. He served as past president of the American Association for Public Opinion Research and the Rural Sociological Society. He has a B.A. in agronomy, an M.S. in rural sociology, and a Ph.D. in sociology, all from Iowa State University.

David Dolson (*Member, Steering Committee*) is director of the Social Survey Methods Division at Statistics Canada, where he is responsible for all statistical and survey methods in support of the Census of Population and National Household Survey, as well as the program of postcensal surveys, the Geography Division, and the population estimates program. He also oversees the Questionnaire Design Resource Centre. He directs the development, testing, evaluation, and implementation of statistical and survey methods, using a variety of data collection modes, including supplementing questionnaire data with information obtained from administrative records. He consulted with the U.S. Census Bureau staff on the Reverse Record Check methodology for census coverage measurement and participated in expert workshops on the U.S. census coverage measurement program and coverage improvement options for the 2020 U.S. census. He has bachelor's and masters of mathematics degrees in statistics from the University of Waterloo.

John L. Eltinge (*Member, Steering Committee*) is the associate commissioner for survey methods research at the U.S. Bureau of Labor Statistics (BLS), where he served previously as a senior mathematical statistician. Prior to working at BLS, he was an associate professor with tenure in the Department of Statistics at Texas A&M University. His primary research interests include survey sampling, alternative data sources, measurement error, incomplete data, survey optimization, survey cost structures, regression trees for complex survey data, and variance function models. He has served as an associate editor for many journals, including *Survey Methodology Journal* and the *Journal of Official Statistics*. In addition, he cochairs the advisory board for the *Journal of Survey Statistics and Methodology* and is a member of the American Statistical Association (ASA) Committee on Fellows and numerous other professional and federal committees. He is a fellow of the ASA and a member of the Federal Committee on Statistical Methodology. He received a B.S. in mathematics from Vanderbilt University, an M.S. in statistics from Purdue University, and a Ph.D. in statistics from Iowa State University.

Scott Fricker (*Presenter*) serves as a senior research psychologist in the Office of Survey Methods Research at the Bureau of Labor Statistics (BLS). His recent research has focused on evaluating measurement error and respondent burden in the Consumer Expenditure Survey, experiments on factors affecting different retrieval strategies in recall surveys, and development and testing of design components for the BLS's new Occupational Requirements Survey. He received a bachelor's degree in psychology from the University of Richmond, a master's degree in social psychology from the University of California, Santa Barbara, and a Ph.D. in survey methodology from the Joint Program in Survey Methodology at the University of Maryland.

Linda Gage (*Cochair, Steering Committee*) retired as senior demographer for the state of California. In this position, her objective was to improve the currency and accuracy of official state and federal demographic data, which were used in policy and funding decisions. She was actively involved in producing and evaluating intercensal population estimates for California and assessing data from the decennial census and the American Community Survey (ACS). She also conducted research commissioned by the U.S. Census Bureau for the ACS 1999-2001 and Census 2000 Comparison Study. She has served for many years on U.S. Department of Commerce and Census Bureau advisory committees and on committees of the Population Association of America. She chaired the three Census Bureau Federal-State steering committees and served as the Governor's Liaison for Census 2000. She has B.A. and M.A. degrees in sociology, with emphasis in demography, from the University of California, Davis.

Jeffrey Gonzalez (*Presenter*) is a research mathematical statistician in the Office of Survey Methods Research at the Bureau of Labor Statistics. His primary research interests include split questionnaire designs, adaptive/responsive designs, total survey error, and statistical computing. He has his Ph.D. in survey methodology (statistical science concentration) in the Joint Program in Survey Methodology at the University of Maryland.

Lauren Harris-Kojetin (*Presenter*) is chief of the Long-Term Care Statistics Branch at the National Center for Health Statistics (NCHS), where she oversees a research program to produce national and state statistical information on the supply, use, and characteristics of providers and users of paid, regulated long-term care services. She has more than 20 years of experience in gerontology, health services research, survey methods, and evaluation, with an emphasis on health care and long-term services and supports for older adults. She is a fellow of the Gerontological Society of America. Before joining NCHS, she directed health services research and survey research projects at LeadingAge and at RTI International. She presents and publishes regularly and serves on several editorial boards. She earned her M.A. and Ph.D. in public policy from Rutgers University.

Steven G. Heeringa (*Presenter*) is a senior research scientist at the University of Michigan Institute for Social Research (ISR). He is a member of the faculty of the University of Michigan Program in Survey Methods and the Joint Program in Survey Methodology. He is a fellow of the American Statistical Association and elected member of the International Statistical Institute. He is the author of many publications on statistical design and sampling methods for research in the fields of public health and the social sciences. He has more than 38 years of statistical sampling experience in the development of the Survey Research Center National Sample design, as well as research designs for ISR's major longitudinal and cross-sectional survey programs. Since 1985, he has collaborated extensively with scientific colleagues in the design and conduct of major studies in aging, psychiatric epidemiology, and physical and mental health. He has a B.S. in biometrics, an M.A. in statistics, and a Ph.D. in biostatistics, all from the University of Michigan.

David Hubble (*Member, Steering Committee*) is a senior statistician at Westat, where he worked on the National Children's Survey, National Assessment of Educational Progress, Minnesota Adult Tobacco Survey, and other survey design and technical assistance projects. Previously, he worked for the U.S. Census Bureau on aspects of designing, planning, and conducting census evaluations and large-scale demographic surveys, including the American Community Survey. His research interests cover a wide range of topics, including survey design, sampling frame creation, sample selection, data collection methods, missing data mitigation, weighting procedures, estimation techniques, variance estimation, methodological investigations, and experimental designs. He has a B.A. and an M.A. in statistics from Boston University.

Frauke Kreuter *(Presenter)* is professor in the Joint Program in Survey Methodology at the University of Maryland and professor of statistics and methodology at the University of Mannheim, Germany. She has additional affiliations with the Maryland Population Research Center, the Institute for Social Research in Michigan, and the German Institute for Employment Research, where she heads the statistical methods group. Prior positions include the Institute for Statistics at the Ludwig-Maximilians University in Munich and the Department of Statistics at the University of California, Los Angeles. Kreuter is a fellow of the American Statistical Association and a recipient of the Gertrude Cox Award from the Washington Statistical Society. Her research focuses on nonresponse errors, paradata and responsive designs, record linkage and, recently, issues of linkage consent, and generalizability for nonprobability samples. She has more than 100 publications, including eight books and monographs. Kreuter was standards chair of the American Association of Public Opinion Research and has served as associate editor or board member for many journals and organizations. She received her B.A. and M.A. in sociology and her Ph.D. in sociology from the University of Konstanz.

Julia I. Lane *(Member, Steering Committee, and Presenter)* is a professor of public service at the New York University (NYU) Wagner Graduate School of Public Service, a professor of practice at the NYU Center for Urban Science and Progress, and an NYU provostial fellow for innovation analytics. Previously, she was a senior managing economist and institute fellow at the American Institutes for Research. In this role, she established the Center for the Science of Science and Innovation Policy and cofounded the Institute for Research on Innovation and Science at the University of Michigan. She has held positions at the National Science Foundation, Urban Institute, World Bank, American University, and NORC at the University of Chicago where she conceptualized and established a data enclave. She also initiated and led the creation and permanent establishment of the Longitudinal Employer-Household Dynamics Program at the U.S. Census Bureau. She has published more than 65 articles and authored and edited eight books. She received a B.A. in economics from Massey University, New Zealand, and a master's degree in statistics and a Ph.D. in economics from the University of Missouri.

Betty Lo (*Presenter*) serves as vice president of community alliances and consumer engagement at Nielsen. In this role, she works with community leaders, media and entertainment companies, and consumer goods companies to promote Nielsen's education, philanthropic, and public affairs efforts to the community, as well as civic and special interest groups. She leads the national strategy for Nielsen's outreach to the Asian American community and partnerships with organizations across the eastern United States. She also leads multicultural advertising efforts. Prior to joining Nielsen, she spent almost 20 years in leading multinational companies, including the Coca-Cola Company and Newell-Rubbermaid. She serves on the national board of the Asian Pacific Islander American (APIA) Chamber of Commerce and Entrepreneurship and the National Association of Asian American Professionals, as well as on the advisory boards for the APIA Scholarship Fund and Organization of Chinese Americans-Asian American Advocates. She has a B.A. in international business from Wesleyan College and an M.B.A. from Emory University.

Nancy A. Mathiowetz (*Member, Steering Committee, and Presenter*) is professor emerita in the Department of Sociology at the University of Wisconsin–Milwaukee (UWM). Prior to joining the faculty at UWM, she was associate professor in the Joint Program in Survey Methodology at the University of Maryland. During her academic career, she taught graduate courses in survey methodology, questionnaire design, statistics, and data analysis. She has published articles on topics related to assessing the quality of survey data, particularly health survey data. She served as editor of *Public Opinion Quarterly* from 2008 to 2012. She is an active member of the American Association for Public Opinion Research (AAPOR), serving as president in 2007-2008; previously she held offices as AAPOR treasurer, standards chair, and membership chair. In 2015, she was awarded the AAPOR Award for Exceptionally Distinguished Service, the association's highest award. She is also an active member of the American Statistical Association and was elected a fellow in 2012. She received a B.S. in sociology from the University of Wisconsin and an M.S. in biostatistics and a Ph.D. in sociology, both from the University of Michigan.

Amy O'Hara (*Presenter*) is chief of the Center for Administrative Records Research and Applications (CARRA) at the U.S. Census Bureau. Her work in CARRA focuses on integrating administrative data into Census Bureau operations and products to reduce respondent burden and data collection costs and improve data quality. She holds a Ph.D. in economics from the University of Notre Dame.

Colm A. O'Muircheartaigh (*Presenter*) is professor and former dean of the University of Chicago's Harris School of Public Policy Studies and a senior fellow at NORC at the University of Chicago. He is an expert in the design and implementation of social investigations. An applied statistician, he has focused his research on the design of complex surveys across a wide range of populations and topics and on fundamental issues of data quality, including the impact of errors in responses to survey questions, cognitive aspects of question wording, and latent variable models for nonresponse. He joined the Harris School faculty in 1998 from the London School of Economics and Political Science, where he was the first director of the Methodology Institute and a faculty member of the Department of Statistics since 1971. A fellow of the Royal Statistical Society and the American Statistical Association and an elected member of the International Statistical Institute, he has served as a consultant to a wide range of public and commercial organizations around the world. He received his undergraduate education at University College Dublin and his Ph.D. in economics from the London School of Economics.

Andy Peytchev (*Presenter*) is a research assistant professor in the Survey Methodology Program at the Institute for Social Research, University of Michigan. He is the principal investigator on research aimed at reducing respondent burden and improving survey estimates through split questionnaire design, by shifting the burden to the survey organization. He also leads the sampling and weighting on a national telephone survey. Previously, he was a senior survey methodologist in the Program for Research in Survey Methodology at RTI International, where he worked on the design and implementation of large-scale government surveys and on methodological investigations. He has a B.S. in marketing from Concord University, an M.A. in survey research and methodology from the University of Nebraska–Lincoln, and a Ph.D. in survey methodology from the University of Michigan.

Elizabeth Poehler (*Presenter*) is a mathematical statistician at the U.S. Census Bureau. She is the chief of the American Community Survey Experiments Branch. She has a B.S. in applied statistics from Rochester Institute of Technology and an M.S. in survey methodology from the University of Maryland, College Park.

Andrew Reamer (*Presenter*) is a research professor at the George Washington Institute of Public Policy. Reamer joined the institute in 2010, after 6 years at the Brookings Institution's Metropolitan Policy Program and 20 years as a consultant in U.S. regional economic development and public policy. His research areas of focus include strategic analysis, innovation, regional economic and workforce development, and the federal economic statistics system. Current and recent project sponsors include the Lumina Foundation, Ewing Marion Kauffman Foundation, Lemelson Foundation, U.S. Census Bureau, the Center for Regional Economic Competitiveness, and the Public Forum Institute. He received a B.S. in economics at The Wharton School, University of Pennsylvania, and a Master of City Planning and Ph.D. in economic development and public policy from the Massachusetts Institute of Technology

Joseph J. Salvo (*Cochair, Steering Committee, and Presenter*) is the director of the Population Division at the New York City Department of City Planning. The division serves as the city's in-house demographic consultant, providing expertise for applications involving assessments of need, program planning and targeting, and policy formulation. He has served on the U.S. Census Bureau's Scientific Advisory Committee and on various panels at the National Academy of Sciences on census issues and is a former president of the Association of Public Data Users. He is coeditor of the *Encyclopedia of the U.S. Census* and coauthor of *The Newest New Yorkers: Characteristics of the City's Foreign-born Population, 2013 Edition.* He is a recipient of the Sloan Public Service Award from the Fund for the City of New York and a fellow of the American Statistical Association. He holds an M.A. and a Ph.D. in sociology from Fordham University.

Deborah Stempowski (*Presenter*) has served as chief of the American Community Survey Office at the U.S. Census Bureau. She began her career at the Census Bureau in 1991 in the Economic Programs Directorate working on the 1992 Economic Census as a data analyst. In 1998, she moved to the Computer-Assisted Survey Research Office and returned to the Economic Directorate in 2005 to lead the effort to implement formal program management practices for the 2007 Economic Census. She also led efforts for company outreach, macro analysis, tabulations, and dissemination operations for the Economic Census. After returning from a detail at the Office of Management and Budget in May 2011, she moved to the director's office. Since April 2012, she has been back in the Economic Directorate and recently became chief of the newly created Economic Management Division. She has a bachelor's degree in economics from Pennsylvania State University and a master's degree in financial management from the University of Maryland, University College.

George Terhanian (*Presenter*) leads The NPD Group's global research sciences, panel management, and analytics and modeling functions. Prior to joining NPD, he was chief strategy and products officer and president, North America, at Toluna. He also spent nearly 14 years at Harris Interactive in leadership positions. He presently serves on the board of directors of the Advertising Research Foundation and the Council of American Survey Research Organizations. He holds a bachelor's degree in political science from Haverford College, a master's degree in education from Harvard University, and a Ph.D. from the University of Pennsylvania.

COMMITTEE ON NATIONAL STATISTICS

The Committee on National Statistics was established in 1972 at the National Academies of Sciences, Engineering, and Medicine to improve the statistical methods and information on which public policy decisions are based. The committee carries out studies, workshops, and other activities to foster better measures and fuller understanding of the economy, the environment, public health, crime, education, immigration, poverty, welfare, and other public policy issues. It also evaluates ongoing statistical programs and tracks the statistical policy and coordinating activities of the federal government, serving a unique role at the intersection of statistics and public policy. The committee's work is supported by a consortium of federal agencies through a National Science Foundation grant.